D0212983

Physics and Chemistry in Space
Volume 9

Edited by
J. G. Roederer, Fairbanks
J. T. Wasson, Los Angeles

Editorial Board:
H. Elsässer, Heidelberg·G. Elwert, Tübingen
L. G. Jacchia, Cambridge, Mass.
J. A. Jacobs, Cambridge, England
N. F. Ness, Greenbelt, Md.·W. Riedler, Graz

A. Nishida

Geomagnetic Diagnosis of the Magnetosphere

With 119 Figures

ANNEX UNIVERSITAIRE BIBLIOTHÈQUE
SCIENTIAE
OTTAVIENSIS
u d'Ottawa
LIBRARY ANNEX

Springer-Verlag

New York Heidelberg Berlin

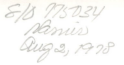

A. NISHIDA

University of Tokyo, Tokyo 153/Japan

The illustrations on the cover are adapted from Figure 29 (right-hand diagram), showing a matrix diagram of the average distribution of equivalent overhead current vectors; and Figure 92 (left-hand diagram), showing examples of plasmapause crossings at different levels of magnetic activity.

ISBN 0-387-08297-2 Springer-Verlag New York Heidelberg Berlin
ISBN 3-540-08297-2 Springer Verlag Berlin Heidelberg New York

All rights reserved.

No part of this book may be translated or reproduced in any form without written permission from Springer-Verlag.

©1978 by Springer-Verlag New York Inc.

The use of registered names, trademarks, etc. in this publication does not imply, even in the absence of a specific statement, that such names are exempt from the relevant protective laws and regulations and therefore free for general use.

Printed in the United States of America

9 8 7 6 5 4 3 2 1

Library of Congress Cataloging in Publication Data. Nishida, Atsuhiro, 1936–. Geomagnetic diagnosis of the magnetosphere. (Physics and chemistry in space; v. 9). Includes bibliographical references and index. 1. Magnetosphere. I. Title. II. Series. QC801.P46 vol. 9 [QC809.M35]523.01'08s ISBN 0-387-08297-2[538'.766]77—21730

Preface

The geomagnetic field observed on the surface of the earth has been an important source of information on the dynamic behavior of the magnetosphere. Because the magnetosphere and its environment are filled with plasma in which electric current can easily flow, dynamic processes that occur in the magnetosphere tend to produce perturbations in the geomagnetic field. Geomagnetic data have therefore provided valuable means for sensing the processes taking place at remote locations, and such basic concepts as the magnetosphere, solar wind, and trapped radiation were derived in early, presatellite days from geomagnetic analyses.

Because of this advantage, geomagnetic observations have been widely utilized for monitoring the overall condition of the magnetosphere. Although the advent of space vehicles has made it possible to observe magnetospheric processes in situ, supplementary information on the overall magnetospheric condition is frequently found to be indispensable for interpreting these observations in the proper perspective. Hence for magnetospheric physicists involved in various branches of the field it has become a common practice to employ geomagnetic data as a basic diagnostic tool.

The main purpose of this monograph is to examine the basis, both experimental and theoretical, of these diagnostic usages of geomagnetic data by discussing geomagnetism in the context of magnetospheric physics in general. Since magnetic variations are such an essential part of magnetospheric physics, the book has in some ways assumed the character of an "introduction to magnetospheric physics" with emphasis on the field aspects. The intended readers are graduate students in space physics and magnetospheric physicists who are not specialists in geomagnetism.

The MKSA unit system is used throughout. In compliance with the conventional practice, however, the unit of $\gamma = 10^{-9}$ Tesla (or, 10^{-9} Wb/m^2) is used to give values of the magnetic induction and the unit of cm^{-3} is used when counting the particle density. Permittivity ε_0 and permeability μ_0 in vacuum are 8.854×10^{-12} Farad/m and $4\pi \times 10^{-7}$ Henry/m, respectively.

It is a pleasure to express gratitude for valuable comments given to the preliminary script of this monograph by Drs. Y.A. Feldstein, T. Iijima, S. Kokubun, L. J. Lanzerotti, K. Maezawa, T. Obayashi, T. Sato, and T. Terasawa. Thanks are due also to Miss E. Suzuki for her diligent typing of the manuscript.

A. NISHIDA

Contents

IV. Dynamic Structure of the Inner Magnetosphere   149

V. Magnetosphere as a Resonator 188

I. Geomagnetic Field Under the Solar Wind

I.1 Introduction: Sudden Commencements and Sudden Impulses

The concept of the magnetosphere grew out of early attempts to understand the cause of geomagnetic sudden impulses.

A geomagnetic sudden impulse (SI) is a sharp change in the geomagnetic field whose onset is recorded within about 1 min all over the world. Sudden impulses can be classified into two groups: positive SIs and negative SIs. A 'positive SI' is characterized by a global increase in the horizontal component H of the geomagnetic field, and a 'negative SI' is characterized by a global decrease in H, although the shape of SI has additional features that depend on both latitude and local time. The rise time of a positive SI and the fall time of a negative SI are usually from one to several minutes. The magnitude of SI is not too large; cases exceeding 50γ ($1\gamma = 10^{-5}$ Gauss or 1 nanotesla) are rather infrequent. Despite their smallness, SIs in ground magnetograms have caught attention by their sharpness and the near-simultaneity of their onset all over the world. The occurrence of an SI is particularly impressive when it precedes an interval of high geomagnetic activity, and such cases of SI (which are generally positive SIs) have been called 'storm sudden commencement (SSC or SC)'.

In their monumental work that laid the foundation of magnetospheric physics, Chapman and Ferraro (1931) conjectured that the SSC is the manifestation of the arrival of a stream of charged particles from the sun. The stream is considered to consist of an equal number of positively and negatively charged particles that are hot enough to yield high electric conductivity. As the stream propagates toward the earth, an electric current is induced on the surface of this highly conductive stream to shield its interior from the geomagnetic field. On the other hand the Lorentz force acting on the shielding current decelerates the stream, and a hollow space is carved out in the solar stream around the earth. Their idea of this hollow space, i.e., the primitive form of the magnetosphere concept, is reproduced in Figure 1. The geomagnetic field inside the hollow space is intensified by the magnetic field produced by the shielding current;

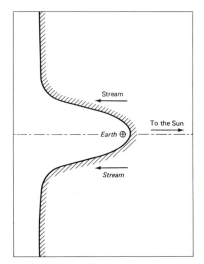

Fig. 1. Early concept of the formation of the magnetosphere (Chapman and Bartels, 1940, slightly modified)

Chapman and Ferraro suggested that this increase constitutes the nature of the SSC phenomenon.

The formation of the magnetosphere and the response of its dimension to changing conditions of the solar stream (now known as the solar wind) have since been pursued as one of the basic topics of magnetospheric physics. In the present chapter we shall first review modern calculations of the formation of the magnetosphere that have been developed in the light of the wealth of data collected by space vehicles since the 1960s. We shall then examine solar-wind and magnetospheric observations during SI events and study detailed mechanisms and consequences of sudden compressions and expansions of the magnetosphere. The chapter is completed by a discussion of the ionospheric current produced in association with SI.

Since no fundamental difference has been detected between morphologies of SI and SSC, we shall treat them together under the term SI. In some literature SI, SSC, and SC have been written in small letters (e.g., si) or by a combination of capital and small letters (e.g., Si).

I.2 Formation of the Magnetosphere

The magnetopause, namely the boundary surface of the magnetosphere, is formed at a distance where the geomagnetic field has become strong enough to stand the compressive force exerted by the solar wind. Earthward of the magnetopause is a regime of the geomagnetic field, while the wide space that lies outside is the realm of the solar wind.

Calculated Shape of the Magnetosphere

The balance of forces that are exerted from both (solar-wind and magnetosphere) sides at the magnetopause has been found to be expressible by a simple pressure-balance equation to a reasonably high degree of approximation:

$$\frac{\mathbf{B}_{\mathrm{M}} \cdot \mathbf{B}_{\mathrm{M}}}{2\mu_0} = Knmv^2 \cos^2 \psi, \tag{1}$$

where $\cos \psi = (\mathbf{v} \cdot \boldsymbol{\xi})/v$. The left-hand side represents the magnetic pressure exerted from jùst inside the magnetopause where the total magnetic field is $\mathbf{B}_{\mathrm{M}}$. The right-hand side represents the dynamic pressure exerted by the solar wind from outside, and K is a factor that depends both on Mach number and on γ (ratio of specific heats) of the solar wind. n and m are solar-wind particle density and mean ion-electron pair mass, $\mathbf{v}$ is the velocity of the upstream solar wind, and ψ is the angle which $\mathbf{v}$ makes with the outward normal $\boldsymbol{\xi}$ to the magnetopause surface.

In this approximation the region within the magnetopause is assumed to be free of plasma, and the presence of the magnetic field in the solar wind, whose effect will be the principal subject of interest in later chapters, is neglected. As far as its contribution to the pressure is concerned, the interplanetary magnetic field $\mathbf{B}_{\mathrm{I}}$ (namely the magnetic field in the solar wind) can be safely neglected since the solar wind is a super-Alfvenic flow satisfying $M_{\mathrm{A}} = v/V_{\mathrm{A}} > 1$, where $V_{\mathrm{A}} = B_{\mathrm{I}}/\sqrt{\mu_0 nm}$, and the dynamic pressure far exceeds the magnetic pressure. The thermal pressure is also more than an order of magnitude less than the dynamic pressure in the solar wind. [Basic parameters of the solar wind are summarized and discussed by Hundhausen (1972).] The simple $\cos^2 \psi$ form to express dependence on the incidence angle may be questioned since the solar wind is actually deflected at the bow shock and does not generally maintain it original direction $\mathbf{v}/v$ until it makes contact with the magnetopause, but this expression has been found to agree reasonably well with the result of detailed gas dynamic calculations (Spreiter et al., 1966).

In terms of the earth-centered polar coordinate system in which the polar axis is directed toward the sun and the dipole axis is in the plane of $\varphi = 90°$, $\boldsymbol{\xi}$ and $\mathbf{v}$ can be expressed as

$$\boldsymbol{\xi} = a\left[\mathbf{r} - \frac{1}{R}\left(\frac{\partial R}{\partial \theta}\right)\boldsymbol{\theta} - \frac{1}{R \sin \theta}\left(\frac{\partial R}{\partial \varphi}\right)\boldsymbol{\varphi} \right] \tag{2}$$

and

$$v = -v \cos \theta\, r + v \sin \theta\, \theta \tag{3}$$

where r, θ, and φ are unit vectors and $R(\theta, \varphi)$ represents the magnetopause surface. a is defined by

$$a = \left[1 + \frac{1}{R^2}\left(\frac{\partial R}{\partial \theta}\right)^2 + \frac{1}{R^2 \sin^2 \theta}\left(\frac{\partial R}{\partial \varphi}\right)^2\right]^{-1/2}. \tag{4}$$

In the final solution of Eq. (1), B_M should be tangential to the boundary surface because the geomagnetic field is supposed to be confined in the magnetosphere. In early stages of the successive approximation to solve the foregoing equation, however, this condition is not met, so that it is customary to substitute $|\xi \times B_M|$, namely the component of B_M parallel to the boundary, for $|B_M|$. The pressure balance Eq. (1) then becomes

$$\left|\left[r - \frac{1}{R}\left(\frac{\partial R}{\partial \theta}\right)\theta - \frac{1}{R \sin \theta}\left(\frac{\partial R}{\partial \varphi}\right)\varphi\right] \times B_M(R)\right|$$

$$= (2\mu_0 K n m v^2)^{1/2}\left(\cos \theta + \frac{\sin \theta}{R}\frac{\partial R}{\partial \theta}\right). \tag{5}$$

In order to solve this equation for $R(\theta, \varphi)$, it is necessary that B_M be known, but B_M in turn is a function of the boundary position expressed by R. Hence a method of successive approximation has to be employed. In the first step of the calculation, B_M is approximated by $2B_G$. Here B_G is the field of the geomagnetic dipole, and 2 is the factor by which the component of the geomagnetic field tangential to the boundary would be intensified if the boundary were planar. When the incidence of the solar wind is perpendicular to the earth's dipole, Eq. (5) reduces to

$$R = r_0 \equiv (2M^2/\mu_0 K n m v^2)^{1/6} \tag{6}$$

at the subsolar point where both $\partial R/\partial \theta$ and $\partial R/\partial \varphi$ vanish due to symmetry. Here B_G at the equator has been expressed as $B_G = M/R^3$, where M is the geomagnetic dipole moment given by 8×10^{15} Wb·m^2 (8×10^{25} Gauss·cm^3). Mead and Beard (1964) have developed an algorism to calculate the surface configuration for the perpendicular incidence case by starting the integration from the subsolar point and by using three-point finite difference formula to represent $\partial R/\partial \theta$ and $\partial R/\partial \varphi$. When solar-wind incidence is not perpendicular to the earth's dipole, the diminished amount of symmetry makes it necessary to solve the pressure balance equation for three points at a time beginning with guesses for R or its derivatives (Olson, 1969; Choe et al., 1973).

The next step of the calculation is to estimate the magnetic field produced by the current flowing on the boundary surface thus derived. Due to the curvature of the boundary the total field $\boldsymbol{B}_\text{M}(\boldsymbol{R})$ at $\boldsymbol{R}$ on the boundary has the contribution $\boldsymbol{B}_\text{c}(\boldsymbol{R})$ from the surface currents $\boldsymbol{J}(\boldsymbol{R}')$ flowing at $\boldsymbol{R}'$ ($\neq \boldsymbol{R}$) where

$$\boldsymbol{B}_\text{c}(\boldsymbol{R}) = \frac{\mu_0}{4\pi} \int_\text{surface} \frac{\boldsymbol{J}(\boldsymbol{R}') \times (\boldsymbol{R}' - \boldsymbol{R})}{|\boldsymbol{R}' - \boldsymbol{R}|^3} \, dS' \tag{7}$$

and

$$\boldsymbol{J}(\boldsymbol{R}') = \frac{2}{\mu_0} \, \boldsymbol{\xi}(\boldsymbol{R}') \times \boldsymbol{B}_\text{G}(\boldsymbol{R}'). \tag{8}$$

[Eq. (8) reflects our basic assumption that there is no magnetic field outside the magnetopause surface.] After calculating $\boldsymbol{B}_\text{c}$, the boundary is determined again by using $2(\boldsymbol{B}_\text{G} + \boldsymbol{B}_\text{c})$ as $\boldsymbol{B}_\text{M}$ in Eq. (5). The solution is made self-consistent by repeating the same procedure but replacing $\boldsymbol{B}_\text{G}$ in Eq. (8) by $(\boldsymbol{B}_\text{G} + \boldsymbol{B}_\text{c})$ with $\boldsymbol{B}_\text{c}$ obtained by the previous approximation.

The calculated shape of the magnetopause is given in Figure 2 for the case of the perpendicular incident solar wind. The noon–midnight

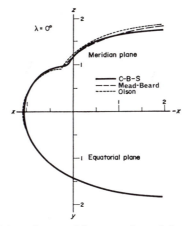

Fig. 2. The noon–midnight and equatorial cross sections of the magnetopause for a solar wind directed perpendicularly to the earth's dipole (Choe et al., 1973, abbreviated as C-B-S). The distance is given in the unit of the first approximation subsolar point distance r_0 and calculations by different authors are compared

meridian section is plotted above the x axis directed toward the sun, and the equatorial cross section is shown below this axis. The radial distance r_b of the magnetopause at the subsolar point (which is frequently referred to as the 'nose' of the magnetosphere) is equal to $1.07r_0$, where r_0 is given by Eq. (6). The equatorial magnetopause distance in the dawn and dusk sections is $1.45r_0$. In the noon–meridian section there is a cusp at the geomagnetic colatitude of about $16°$. In the present idealization, in which the interplanetary magnetic field is neglected, there exists in this cusp a magnetic neutral point. At the neutral point in the southern hemisphere the geomagnetic field lines emerge from the interior of the magnetosphere and spread over the entire magnetopause, whereas at the northern neutral point the field lines on the magnetopause converge and enter the magneto-sphere interior.

The dependence of the magnetospheric configuration on the solar-wind incidence direction is illustrated in Figure 3. λ is the angle which the incidence direction makes with the geomagnetic equatorial plane. It is seen that the attitude of the dayside magnetosphere is strongly tied to the dipole direction, but the night-side tail of the magnetosphere tends to be aligned to the solar-wind incidence direction. As a result, the calculated cusp position shows a strong dependence on λ; the geomagnetic colatitude

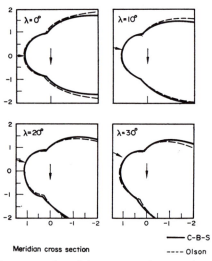

Fig. 3. Noon–midnight cross section of the magnetosphere as a function of the solar-wind incidence angle λ (Choe et al., 1973; Olson, 1969). The distance is in the unit of r_0, and directions of the solar wind and of the geomagnetic dipole are indicated by arrows

of the northern neutral point varies from $12°$ to $28°$ as λ swings from $+35°$ to $-35°$ according to Choe et al. (1973).

The magnetic field produced inside the magnetosphere by the magnetopause current has been expressed by Choe and Beard (1974) in the form of the spherical harmonic expansion:

$$B_s(r, \theta, \varphi, \lambda) = -\nabla\varphi_s(r, \theta, \varphi, \lambda)$$

$$\varphi_s(r, \theta, \varphi, \lambda) = \sum_{l=1} r^l \sum_{m=0}^{l} g_{lm}(\lambda) P_{lm}(\cos \theta) \cos m\varphi \tag{9}$$

where $g_{lm}(\lambda)$ is a power series in λ. The spherical coordinates θ and φ now refer to geomagnetic colatitude[1] and local time. At the earth's surface the lowest order term ($l = 1$) is predominant, and at the equator B_s is directed northward in the case of the perpendicular incidence. The local time average of the perturbation field at the equator is given roughly by

$$\left| \left\langle B_s \left(R_E, \frac{\pi}{2}, \varphi, 0 \right) \right\rangle_\varphi \right| = 0.20 \times 10^5 \left(\frac{r_0}{R_E} \right)^{-3}$$

$$= 1.6 \times 10^5 \sqrt{10 K n m v^2} \tag{10}$$

in the unit of γ, where $\langle \ \rangle_\varphi$ denotes average over φ (local time) and n, m, and v are in MKS units.

The value of the factor K, which applies to the interaction of a blunt body with a supersonic stream, is about 0.8 (Spreiter et al., 1966). Using this value for K we can present the following as numerical examples. When the solar wind has a velocity of 300 km/s and consists only of proton-electron pairs with a density of $5/\text{cm}^3$, the magnetopause distance r_b at the subsolar point is $12.5 R_E$ and the average equatorial field produced on the ground by the magnetopause current is 12.5γ. The increase of solar-wind velocity and density to 600 km/s and $15/\text{cm}^3$ brings r_b to $8.2 R_E$, and the surface field produced by the magnetopause current is enhanced to 43γ. $\lambda = 0$ is assumed and the earth's induction is not considered in the preceding estimates.

---

[1] The geomagnetic coordinate system is a spherical coordinate system whose origin is the center of the earth and whose axis is parallel to the dipole moment of the earth's main field. Geomagnetic longitude is measured eastward from the American side of a plane that contains both the earth's rotational axis and the geomagnetic axis.

Throughout this monograph the coordinate shown in parentheses following a station name is the geomagnetic latitude of the station, except when noted otherwise.

Observed Shape of the Magnetosphere

Observations of magnetopause positions are summarized and compared with the theoretical calculation in Figure 4. The Figure represents the section of the magnetopause at the solar ecliptic plane, and 474 magnetopause crossings are plotted by small crosses. Since actual crossings are not made in general in the ecliptic plane, they are projected to the ecliptic by the rotation in a meridian plane (for day-side observations) or about the earth–sun line (for night-side observations). The coordinates X'_{SE} and Y'_{SE} refer to the solar ecliptic coordinate system[2] rotated westward by 4°. The westward rotation of 4° is to take into account the aberration of the solar-wind direction due to the earth's orbital motion. The solid curve represents, at $X'_{SE} > -15R_E$, a best-fit ellipse to the observed magnetopause locations. At $X'_{SE} \leq -15R_E$ it is a straight-line extension from this ellipse.

Observations are not normalized to a standard value of the solar-wind momentum flux, and the theoretical magnetopause configuration shown by a dashed curve in Figure 4 is the equatorial section of a solution corresponding to $\lambda = 0$ and $r_b = 10.7R_E$ (Olson, 1969). (Magnetopause configurations obtained by different authors under the $\lambda = 0$ condition differ little in the equatorial section.) Calculated and observed magnetopause configurations agree fairly well on the dayside, but from the dawn–dusk section to the tail the observed dimension is systematically larger than the calculated one. This is so despite the fact that the neglected terms in the solar-wind pressure, namely the thermal and magnetic pressures, tend to become more important at flanks and tail surfaces where $\cos^2 \psi$ is small. The preceding discrepancy therefore suggests that the pressure exerted from inside the magnetopause is not simply the one that originates from the geomagnetic dipole field.

A quantitative comparison has also been carried out between observed and predicted sizes of the magnetosphere using solar-wind parameters obtained immediately before, or after, magnetopause crossings. Observed sizes are $2 \sim 3\%$ larger than the predicted, and the presence of an additional internal pressure has been deduced from this point too (Fairfield, 1971).

---

[2] In the solar ecliptic coordinate system, X_{SE} is along the earth–sun line and positive in the sunward direction, Z_{SE} is perpendicular to the ecliptic plane and positive in the northward direction, and Y_{SE} completes the right-handed orthogonal system. Solar ecliptic colatitude is measured from the Z_{SE} axis, and longitude is measured from the $X_{SE} - Z_{SE}$ plane counterclockwise.

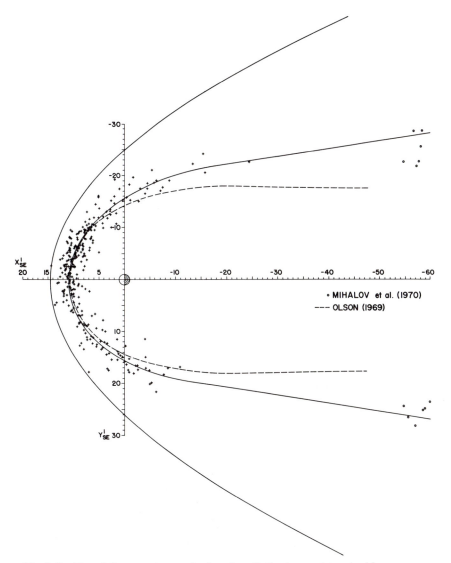

Fig. 4. Position of the magnetopause in the solar ecliptic plane as determined by measurements on six IMP spacecrafts, 1963–1968 (Fairfield, 1971). (+, •) The average location observed on individual passes. *Inner solid curve*: the best-fit curve to these points. (––––) The theoretical magnetopause configuration at the equator obtained by Olson (1969). *Outer solid curve*: the best-fit hyperbola to bow shock crossing positions

The source of the additional pressure should be the plasma residing in the magnetosphere. In addition to exerting the thermal pressure, the resident plasma influences the pressure balance condition by generating additional magnetic pressure that originates from the current it carries. Since the plasma is placed in the magnetic field, its drift motion produces current j_P, perpendicular to the magnetic field, given by

$$j_P = \frac{B}{B^2} \times \nabla P_\perp + \frac{P_\parallel - P_\perp}{B^4} B \times [(B \cdot \nabla)B] \tag{11}$$

where $P_\perp$ and $P_\parallel$ are perpendicular and parallel pressures, respectively, relative to the magnetic field. The current can also flow along the magnetic field if there is a suitable driving mechanism. What has particularly important influence on the configuration of the night-side magnetosphere is the drift current flowing in the plasma sheet. As illustrated in Figure 5 the plasma sheet is a region of warm plasma that occupies the midplane of the magnetospheric tail (sometimes referred to as magnetotail) with a thickness of several R_E. In the near-earth region of $|x| \lesssim 25R_E$ the electrons in the plasma sheet are characterized by number densities of 0.3 to 30/cm^3 and mean energies of 50 to 1600 eV. There is strong anticorrelation between density and mean energy, so that the electron energy density and energy flux usually stay around ~ 1 keV/cm^3 and ~ 3 erg/cm^2·s, respectively (Vasyliunas, 1968). The energy density of protons is up to an order of magnitude higher than that of electrons (Frank, 1971). At the

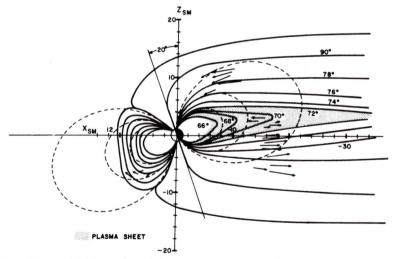

Fig. 5. Noon–midnight section of the magnetosphere. *Dotted region*: the plasma sheet. (———) Dipole field lines, for reference (Fairfield and Ness, 1970)

lunar distance of $|x| \sim 60R_E$ typical quiet-time parameters are: density $\sim 0.10/cm^3$, electron energy ~ 0.20 keV, proton energy ~ 2.5 keV, and plasma pressure $\sim 4.3 \times 10^{-11}$ Newton/m^2 (Rich et al., 1973). The magnetotail has been found to extend as far as $1000R_E$ away from the earth (Walker et al., 1975).

Since the plasma sheet represents a thin layer whose thickness is much smaller than its width and length, its basic structure can be described in terms of a one-dimensional model that depends only on the coordinate z across the sheet. Hence $\partial B_x/\partial z = \mu_0 j_P$. When the magnetic field is assumed to consist only of the component B_x parallel to the sheet, the y component of Eq. (11) reduces to

$$\frac{\partial}{\partial z}\left(\frac{B_x^2}{2\mu_0} + P_\perp\right) = 0, \tag{12}$$

which means that the sum of magnetic and plasma pressures is kept constant across the sheet. Eq. (12) can be utilized to predict the strength of the magnetic field in the high latitude lobe region where $P_\perp \doteq 0$; at the lunar distance the pressure observed in the plasma sheet predicts the lobe B_x of about 10γ. The field of this magnitude has indeed been detected by magnetic field observations (Behannon, 1970). Thus at that distance the field due to the plasma sheet current is much stronger than the dipole field ($\sim 0.1\gamma$), and the structure of the magnetosphere is governed by this additional field. [The spatial variation of the tail field strength with distance from the earth has been found to be expressible as $|B| \propto |x|^{-0.3}$ (Behannon, 1970).]

The position of the plasma sheet relative to the geomagnetic equatorial plane varies with the incidence angle λ of the solar wind. Since the attitude of the magnetotail depends both on the dipole orientation and the solar-wind direction, the solar magnetospheric coordinate system[3] is the relevant framework for describing this variation. From particle and field observations in the magnetotail the distance Δz of the tail midplane from the solar magnetospheric equatorial plane has been found to be expressible as

$$\Delta z = \begin{cases} R_0 \sin \lambda & \text{at } |y| \le Y_0 \\ 0 & |y| > Y_0 \end{cases} \tag{13a}$$

---

[3] In the solar magnetospheric coordinate system, X_{SM} is along the earth–sun line and positive in the sunward direction, Z_{SM} is in the plane including both X_{SM} axis and the geomagnetic dipole moment and positive in the northward direction, and Y_{SM} completes the right-handed system. Solar magnetospheric colatitude is measured from the Z_{SM} axis, and longitude is measured counterclockwise from the $X_{SM} - Z_{SM}$ plane.

where $R_0 = Y_0 = 8R_E$ (Murayama, 1966). The formula is revised later to

$$\Delta z = \begin{cases} (R_0^2 - y^2)^{1/2} \sin \lambda & \text{at } |y| \leq Y_0 \\ 0 & \text{at } |y| > Y_0 \end{cases} \tag{13b}$$

where $R_0 = Y_0 = 11R_E$ (Russell and Brody, 1967), or,

$$\Delta z = \begin{cases} \left(R_0^2 - \dfrac{R_0^2}{Y_0^2} y^2 \right)^{1/2} \sin \lambda & \text{at } |y| \leq Y_0 \\ 0 & \text{at } |y| > Y_0 \end{cases} \tag{13c}$$

where $R_0 = 11R_E$ and $Y_0 = 15R_E$ (Fairfield and Ness, 1970). The slight difference among these three forms of the Δz formula lies in the way the y dependence is expressed, and in all cases the dependence on the x coordinate is assumed to be insignificant beyond $|x| \sim R_0$. However, at much greater distances ($|x| \gtrsim 30R_E$) it becomes difficult to express Δz by a simple formula since irregular flapping motions tend to dominate the systematic variation dependent on the λ angle.

In Figure 4, presented earlier, the form of the earth's bow shock was also depicted. It was represented by a best-fit hyperbola to 389 observed shock crossing positions. The bow shock is formed because the solar-wind speed relative to the earth is supersonic and super-Alfvenic, and the observation of the sharp, well-defined shock front is a decisive demonstration of the continuum character of the solar-wind plasma. The observed ratio of the stand-off distance to the magnetopause distance is 0.33, and this indicates that the effective value of the ratio γ of specific heats in the solar wind is between 5/3 and 2 (Fairfield, 1971). The region between the bow shock and the magnetopause is called magnetosheath.

I.3 Sudden Changes in the Magnetospheric Dimension

Since the solar wind is by no means a uniform and steady flow, the pressure exerted on the magnetopause changes from time to time, and the dimension of the magnetosphere changes accordingly. In particular, upon arrival of sharply defined solar-wind structures the magnetosphere is suddenly compressed or extended, and sudden impulses are produced. This effect is demonstrated in Figure 6 where simultaneous observations of the solar-wind momentum flux (expressed as nv^2) and the low-latitude geomagnetic field are compared. In the March 19, 1971 case (upper panel) there is a clear correspondence between an enhancement in nv^2 of the

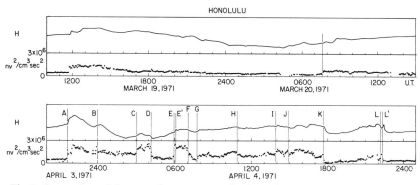

Fig. 6. Comparison of the ground magnetogram (horizontal component at Honolulu) with simultaneous solar-wind data (Ogilvie and Burlaga, 1974).

solar wind and a positive SI around 1145, and in the April 3 to 4 case (lower panel), many correspondences (labeled A through L′) between nv^2 changes and positive/negative SIs are noted.

Waves and Discontinuities in Magnetized Plasma

In order to understand the nature of sudden changes in the state of the solar wind as observed above we have to consult basic conservation equations in magnetohydrodynamics. Let us adopt a frame of reference that moves with the surface where the state of the solar wind sharply changes, and assume that the flow is steady relative to this frame of reference. We shall also assume that pressure is sufficiently isotropic and expressible as a scalar. The conservation of *mass* can be written as

$$[\rho v_n] = 0 \tag{14}$$

where ρ and v refer to mass density and velocity, and the subscript n refers to a component normal to that surface. Brackets $[\ ]$ are defined to mean $[Q] = Q_1 - Q_0$ where subscripts 0 and 1 refer to conditions on the upstream and downstream sides of the discontinuity. The conservation of *momentum* is expressed as

$$\left[\rho v_n v + \left(p + \frac{B^2}{2\mu_0} \right) n - \frac{1}{\mu_0} B_n B \right] = 0 \tag{15}$$

where p and B refer to pressure and magnetic field, n refers to unit vector normal to the surface, and subscript t will define the component tangential to the surface. The conservation of *energy* is expressed as

$$\left[\rho v_n (\tfrac{1}{2} v^2 + h) + v_n \frac{B^2}{\mu_0} - \frac{1}{\mu_0} B_n \boldsymbol{v} \cdot \boldsymbol{B} \right] = 0 \tag{16}$$

where h refers to enthalpy. The conservation of *magnetic flux* is expressed as

$$[B_n] = 0. \tag{17}$$

In addition, since the electric field E satisfies rot $E = 0$ in the steady state, we have $[E_t] = 0$, so that by using the 'frozen-in' relation ($E + v \times B = 0$) we have

$$[B_n v_t - B_t v_n] = 0. \tag{18}$$

The solution of the foregoing equations can be classified into two modes according to whether or not $v_n = 0$. When $v_n = 0$ there is no net flow of plasma across the surface; the discontinuity is attached to the flow and convected with it. Such a discontinuity is designated as tangential or contact discontinuity. When the magnetic field is parallel to the discontinuity, namely if

$$v_n = B_n = 0 \tag{19}$$

Eq. (15) becomes

$$\left[p + \frac{B^2}{2\mu_0} \right] = 0, \tag{20}$$

which means that the total pressure across the discontinuity is in balance. The discontinuity of this kind is called *tangential discontinuity*. Other conditions [Eqs. (14)–(18)] are automatically satisfied. Eq. (20) requires that at a tangential discontinuity changes in n, T, and B cannot all have the same sign. Tangential discontinuities have been found to be abundant in the solar wind (Nishida, 1966a; Burlaga, 1971). On the other hand, when $v_n = 0$ but $B_n \neq 0$, namely, when there is a component of the magnetic field across the discontinuity, Eqs. (15)–(18) yield

$$[B_t] = 0, \qquad \text{hence } [B] = 0 \tag{21}$$

and

$$[v_t] = 0, \qquad \text{hence } [v] = 0 \tag{22}$$

and from Eqs. (15) and (21) it also follows that

$$[p] = 0. \tag{23}$$

Such a discontinuity is called *contact discontinuity*. Contact discontinuities are not likely to represent a stable structure because plasma on both sides would migrate by diffusion along field lines and smooth out the discontinuity.

Another mode of the solution to the conservation equations is characterized by $v_n \neq 0$, which means that there is flow of plasma across the surface. In other words, the structure propagates through the medium. Solutions of this mode therefore represent waves, and since sharp, large-amplitude changes are considered here, they are classified as shock waves. To study shock solutions it is more convenient to rewrite Eqs. (14)–(18) in alternative forms introducing variables $\tau = 1/\rho$ and $F = \rho v_n$ and using mean values $\langle Q \rangle = (Q_0 + Q_1)/2$. First, it immediately follows from Eq. (14) that

$$F[\tau] - [v_n] = 0. \tag{24}$$

We also have

$$F[\boldsymbol{v}] + [p]\boldsymbol{n} + \frac{1}{\mu_0} \langle \boldsymbol{B} \rangle \cdot [\boldsymbol{B}]\boldsymbol{n} - \frac{1}{\mu_0} B_n[\boldsymbol{B}] = 0 \tag{25}$$

$$F\left\{ [e + \langle p \rangle \tau] + \frac{1}{4\mu_0}[\tau][\boldsymbol{B}_t]^2 \right\} = 0. \tag{26}$$

These are transformations of Eqs. (15) and (16), and $e = h - p/\rho$ is the internal energy. We have in addition

$$F \langle \tau \rangle [\boldsymbol{B}] + \langle \boldsymbol{B} \rangle [v_n] - B_n[\boldsymbol{v}] = 0. \tag{27}$$

The t component of this equation is identical to Eq. (18) and its n component is justified by Eq. (17).

In the rewritten form the basic equations become a set of linear equations. From Eqs. (24)–(27), the equation for the mass flux F can be obtained:

$$\langle \tau \rangle^2 F\left(\langle \tau \rangle F^2 - \frac{B_n^2}{\mu_0} \right) \left\{ \langle \tau \rangle F^4 + \left(\frac{\langle \tau \rangle}{[\tau]}[p] - \frac{\langle B \rangle^2}{\mu_0} \right) F^2 - \frac{[p]B_n^2}{[\tau]\mu_0} \right\} = 0. \tag{28}$$

$F = 0$ corresponds to discontinuity solutions discussed earlier, and shock solutions corresponds to $F \neq 0$. Shocks associated with the roots

$$F = \pm B_n/(\mu_0 \langle \tau \rangle)^{1/2} \tag{29}$$

are called *intermediate shocks* (or, Alfven shocks, transverse shocks, or rotational discontinuities). Combining t components of Eqs. (25) and (27) and using Eq. (29) to eliminate F, one obtains

$$[v_n] = 0 \qquad (30)$$

and hence

$$[\tau] = 0, \qquad (31)$$

namely, intermediate shocks do not cause change in density. The velocity of propagation of intermediate shocks with respect to the medium, $-v_n$, is given by $V_{A,n}$ where

$$V_{A,n} = \pm B_n/(\mu_0 \rho)^{1/2}, \qquad (32)$$

which is the Alfven speed corresponding to the normal component of the magnetic field. In the solar wind, velocity changes associated with intermediate shocks are small; usually their magnitudes are $\lesssim 10\%$ of the velocity (Belcher and Davis, 1971). Hence dimensional changes of the magnetosphere due to intermediate shocks are not expected to be very large.

Other shock solutions arise when the last factor of Eq. (28) is equated to zero. This condition can be written as

$$\left(F^2 + \frac{[p]}{[\tau]} \right) \left(\langle \tau \rangle F^2 - \frac{B_n^2}{\mu_0} \right) = F^2 \frac{\langle B_t \rangle^2}{\mu_0}. \qquad (33)$$

Since the right-hand side is always positive, two roots F_f^2 and F_s^2, where $F_f^2 > F_s^2$, should satisfy

$$F_f^2 > -\frac{[p]}{[\tau]} \text{ and } F_A^2 \qquad (34)$$

$$F_s^2 < -\frac{[p]}{[\tau]} \text{ and } F_A^2 \qquad (35)$$

where $F_A = \pm B_n/(\mu_0 \langle \tau \rangle)^{1/2}$ is the root corresponding to the intermediate shock solution. Shock waves characterized by F_f are called *fast shocks*, and those by F_s *slow shocks*. Within the shock front the kinetic energy of the bulk motion of the solar wind is dissipated and the entropy is increased. Therefore jumps in pressure and density across the shock should satisfy

$$[p] > 0 \text{ and } [\tau] < 0. \tag{36}$$

From t components of Eqs. (25) and (27) it follows that

$$[B_t] = \frac{-F^2[\tau]\langle B_t \rangle}{\langle \tau \rangle F^2 - B_n^2/\mu_0} \tag{37}$$

and hence

$$[B_t^2] = 2\langle B_t \rangle \cdot [B_t] = -\frac{2F^2[\tau]\langle B_t \rangle^2}{\langle \tau \rangle F^2 - B_n^2/\mu_0}. \tag{38}$$

The sign of the denominator on the right-hand side of this equation depends on the mode of the shock. For fast shocks it is positive due to (34) and we have

$$[B_t^2] > 0, \quad \text{hence } [B^2] > 0, \tag{39}$$

namely, the strength of the magnetic field increases through a fast shock wave. For slow shocks it follows from (35) that

$$[B_t^2] < 0, \quad \text{hence } [B^2] < 0, \tag{40}$$

namely, the magnetic field is weakened through a slow shock wave. For shocks of both modes it can be proved that the B vector on both sides of the shock should lie on a plane that contains the shock normal; this property is known as the coplanarity theorem (Spreiter et al., 1966; Colburn and Sonett, 1966).

Jump conditions at discontinuities and shock waves are summarized in Table 1. Here an asterisk means that value of the jump of the given parameter is arbitrary but not zero.

Table 1. Classification of shock waves and discontinuities

	v_n	B_n	$[v]$	$[B]$	$[B^2]$	$[p]$	$\left[p + \dfrac{B^2}{2\mu_0}\right]$	$[\rho]$
Tangential discontinuity	0	0					0	
Contact discontinuity	0	*	0	0	0	0	0	*
Intermediate shock	*	*	*	*	0	0	0	0
Fast shock	*		*	*	+	+	+	+
Slow shock	*	*	*	*	−	+	+	+

[ ]: jump in the given variable

Origin of Sudden Impulses

Detailed structures of the solar wind that gave rise to sudden impulses of Figure 6 can be seen in Figures 7a and 7b. Each figure displays, from the top, records of energy flux ($\frac{1}{2}nv^3$) of the bulk motion, bulk velocity (v), proton density (n_p), proton temperature (T_p), and flux density (B), latitudinal angle (θ), and longitudinal angle (φ) of the interplanetary magnetic field in the solar ecliptic coordinate system. The last record is the ratio between

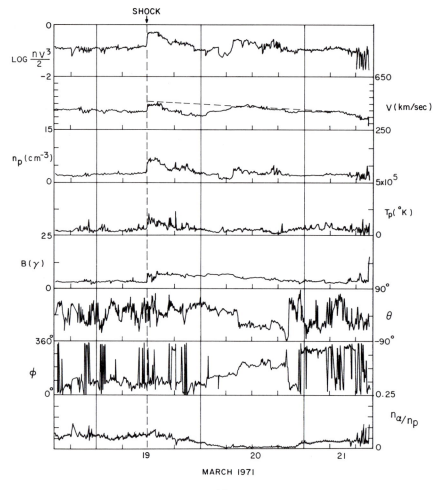

(a)

Fig. 7a,b. (a) Detailed solar-wind data corresponding to the upper panel of Figure 6, and (b) to the lower panel of Figure 6 (Ogilvie and Burlaga, 1974)

α-particle density (n_α) and proton density. Figure 7a corresponds to the upper panel of Figure 6 and shows that the increase in the dynamic pressure at 1145 on March 19 is the result of simultaneous increases in n_p, T_p, and B; namely, increase in ρ, p, B^2, and $p + B^2/2\mu_0$. Referring to Table 1, we can identify this increase with a fast shock. The simultaneous increase in the bulk velocity is consistent with this interpretation. A minor solar flare that took place about 53 h before has been suggested as the origin of this fast shock (Ogilvie and Burlaga, 1974).

Figure 7b corresponds to the lower panel of Figure 6, where numerous cases of changes in nv^2 have been recognized. The first of these

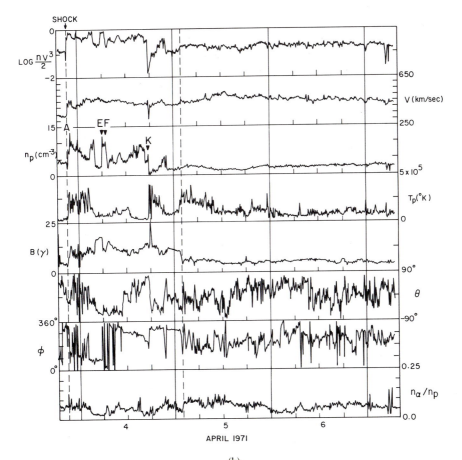

(b)

Fig. 7 (*cont.*)

increases, labeled as event A, can be seen to be due to simultaneous increases in v, n_p, T_p, and B. Hence the structure is identified as a fast shock, and it probably originated from a flare of importance $3N$ that took place 59 h before. Other increases in nv^2 seem to be due to tangential discontinuities. For example, the E event at ~ 0600 April 4 corresponds to an increase in n_p and a decrease in B with v and T_p remaining almost unaffected. Although a rigorous examination of the pressure balance condition $[p + B^2/2\mu_0] = 0$ cannot be made because the electron temperature data are not available, the necessary condition for satisfying the foregoing condition is met, and referring to Table 1, we note that only the tangential discontinuity can cause the observed combination of behaviors in solar-wind parameters. The event F that follows is an opposite of the E event; n_p decreases and B increases, while T_p is not affected. This obviously is another case of the tangential discontinuity. Frequently, two tangential discontinuities having opposite characteristics occur in close succession, as in this instance, reflecting the presence of the filamentary structure in the solar wind. At event K around 1800 April 4, an increase in T_p is observed simultaneously with a decrease in n_p, so that the event can be identified with a tangential discontinuity. A brief negative spike in T_p that immediately follows the above increase is accompanied by a sharp positive spike in B while n_p is almost unaffected; this also seems to represent a pair of tangential discontinuities.

It should be noted that of the numerous cases of SIs observed on April 3 to 4, 1971, only the event A is tabulated as SSC or SI in the "Sudden commencements and solar flare effects" Section of the *Solar Geophysical Data*. Although large and isolated SI events resulting from the shock waves have a relatively good chance of being identified as SSCs or SIs by geomagnetic observatories, smaller and repetitive events like events B through L′ above are frequently missed; solar-wind structures are much more prevalent than might be inferred from the available SSC and SI listings (Nishida and Jacobs, 1962a; Burlaga, 1971).

In addition to the fast shocks and tangential discontinuities exemplified above, cases of slow shocks have been identified in the solar wind (Chao and Olbert, 1970). Moreover, detections have also been made of those shock waves (both fast and slow) that propagate backward against the solar wind (Dryer, 1975). These reverse shocks are considered to be produced in high-speed solar streams or clouds as they collide with the foregoing, slower flow and are decelerated (Colburn and Sonett, 1966).

The relation between changes in the solar-wind dynamic pressure and in the ground magnetic field is examined quantitatively in Figure 8. According to Eq. (10) we expect that the magnitude I_{SI} of a sudden impulse in low latitudes averaged over local times is related to the change

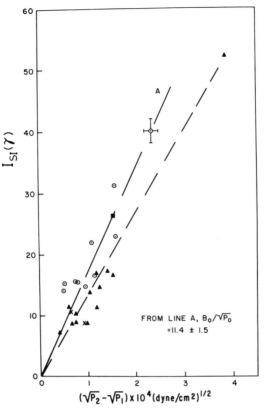

Fig. 8. Comparison of the local-time averaged magnitude of sudden impulses in low latitudes with corresponding change in the dynamic pressure of the solar wind. P_1 and P_2: dynamic pressures on upstream and downstream sides of the solar-wind structure, respectively (—— Ogilvie et al., 1968; ---- Siscoe et al., 1968)

in $\sqrt{Knmv^2}$ as follows;

$$I_{SI} = 1.5 \times 1.6 \times 10^5 \, \varDelta\sqrt{10Knmv^2} \qquad (41)$$

where nm should be read as the total density of the solar-wind plasma including the α-particle contribution, and the magnetic field is in γ and n, m, and v are in MKS units. A factor of 1.5 has been included to take into account the effect of the earth's induction. Figure 8 confirms the linear dependence of SI magnitudes on changes in the solar-wind dynamic pressure but reveals appreciable differences in the values for the proportionality constant. The empirical relation obtained from Figure 8 is

$$I_{SI} = 1.14 \times 10^5 \, \varDelta\sqrt{10nmv^2} \qquad (42)$$

and hence the observed SI field is about two times weaker than the expected one. [To compare Figure 8 with Eq. (42) note that pressure given in dyn/cm^2 is 10 times that expressed in Newton/m^2.]

It has been suggested that this difference is due to the magnetic field produced by energetic particles trapped in the radiation belt (Siscoe et al., 1968). Drift of these particles in the magnetic field produces so-called ring current given by Eq. (11), and the resulting magnetic field adds to the geomagnetic field at the magnetopause but subtracts from it at the earth's surface. As will be discussed in Chapter IV, the magnitude I_{DR} of the field produced on the ground by the ring current is related to the total kinetic energy content U_T of the trapped particle population by

$$ I_{DR} = -\frac{2}{3} \frac{U_T}{U_D} B_D \tag{43} $$

where B_D is the dipole field strength on the ground at the equator and U_D is the energy in the dipole magnetic field above the earth's surface ($\sim 9 \times 10^{17}$ Joules). When the magnetosphere contracts or expands, the strength of the magnetic field increases or decreases throughout the magnetosphere, and the kinetic energy of individual particles residing in the magnetosphere is enhanced or reduced accordingly, due to the conservation of first and second adiabatic invariants. Resulting change in U_T causes change in I_{DR} that acts to reduce the field I_{SI} produced by the magnetopause current. The ring-current field produced on the magnetopause side, on the other hand, acts to resist the change in the magnetospheric dimension, and would also help to lower the magnitude of I_{SI}.

I.4 Hydromagnetic Propagation of the SI Signal

Sudden deformations of the magnetosphere initiated at the magnetopause spread throughout the magnetosphere and reach the ground as hydromagnetic waves if their time scale is much longer than the period of ion gyration. Magnetospheric deformation associated with SI events fall in this category, since the period of the ion gyration in the magnetosphere is about 1 s or less and is much shorter than SI time scales.

SI as Propagating Fast-Mode Wave

While the magnetospheric compression or expansion is in progress and the magnetopause is moving with speed v_b, the pressure balance

Eq. (1) at the subsolar point has to be replaced by

$$\frac{B^2}{2\mu_0} = K n_s m_s (v_s - v_b)^2 \qquad (44)$$

where B is the instantaneous value of the magnetic field immediately inside the magnetopause and subscript s has been used to specify solar-wind density n, speed v, and mean ion-electron pair mass m. The motion continues until B reaches the final, equilibrium value B_M.

$$\frac{B_M^2}{2\mu_0} = K n_s m_s v_s^2. \qquad (45)$$

Hence v_b can be written as

$$v_b = \frac{B_M - B}{\sqrt{2\mu_0 K n_s m_s}} \qquad (46)$$

where positive v_b means earthward motion. The compression or expansion initiated at the magnetopause spreads throughout the magnetosphere with speed V_F of the hydromagnetic wave of the fast mode:

$$V_F = \left\{ \left(\frac{\gamma p_m}{n_m m_m} \right) + \left(\frac{B_m}{\sqrt{\mu_0 n_m m_m}} \right)^2 \right\}^{1/2} \qquad (47)$$

where subscript m specifies magnetospheric variables. The motion of the magnetopause is subsonic, in the sense that $v_b / V_F < 1$, if $n_m \ll n_s$ and $m_m \approx m_s$.

The distribution of V_F with radial distance is illustrated in Figure 9 for the daytime condition. Since the ratio β between thermal energy

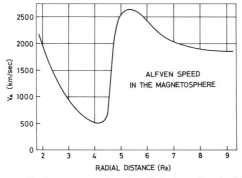

Fig. 9. Alfven wave speed in the magnetosphere for relatively undisturbed daytime conditions

density and magnetic energy density is less than one in the day-side magnetosphere, V_F is practically equal to Alfven speed V_A:

$$V_A = \frac{B_m}{\sqrt{\mu_0 n_m m_m}}.$$

(48)

In the present estimate of V_A we have used particle and field information
corresponding to relatively undisturbed conditions; the field is after Choe
and Beard's (1974) model corresponding to the subsolar magnetopause
distance of $10.7R_E$, and plasma density is after Chappell et al.'s (1970)
observation on March 12, 1968. The contribution of energetic particles
to density is neglected and the field distortion by the ring current is not
taken into account.

With the speed given by Figure 9 the transmission of the signal from
the magnetopause to the surface of the earth would occur within about
one minute. This expectation has been confirmed by comparative studies
of ground and magnetospheric observations of SIs such as those shown
in Figure 10, where (a) is for a positive SI and (b) is for a negative SI. In
each figure, radial, latitudinal, and azimuthal components of the magnetic
field observed by a satellite are displayed in the first three panels, while
the magnitude of the field is given in the fourth panel. All these quantities
are corrected for the contribution from the geomagnetic main field by
the Jensen and Cain (1962) model. The last panel shows ground magnetograms obtained at Guam (geomag. lat. 4°) in the order of D (declination),
H (horizontal component), and V (vertical component). It can be seen
that positive or negative SI is observed on the ground within one minute
of the detection of the field increase or decrease at the satellite (which was
located near the noon meridian at the radial distance of $4 \sim 6R_E$). Since
the propagation speed of the signal in the outer magnetosphere is greater
than the propagation speed of the solar-wind structure (which exceeds
1000 km/s only in rare occasions), in one reported instance the magnetic
field inside the magnetotail started to increase well before the arrival of
the interplanetary shock wave at the tail surface closest to the observing
site; the compressional wave that was generated on the dayside magnetopause arrived earlier via the magnetospheric path (Sugiura et al., 1968).

The ratio between magnitudes of SIs observed in the magnetosphere
by satellites to that observed on the ground is expected to lie between 1
and about 2, according to model calculations of the field produced by
the magnetopause current (e.g., Choe and Beard, 1974). This ratio has,
however, been observed to deviate from the foregoing range (Patel and
Coleman, 1970). The discrepancy is probably due to the modulation of

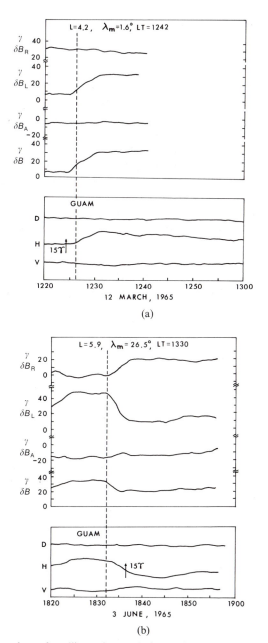

Fig. 10a,b. Comparison of satellite and ground observations of (a) a positive SI and (b) a negative SI (Patel and Cahill, 1974). Satellite positions at the time of the observation are indicated at the top

the ring current field by magnetospheric compression or expansion as discussed at the end of the last section. In addition, modifications of the ionospheric current during SIs should also contribute to this discrepancy.

The rise time of positive SIs in low latitudes on the ground has been found to lie between 1 to 6 min in most cases, and it has been found to show loose anticorrelation with the SI amplitude (e.g., Mayaud, 1975). The fall time of negative SIs also lies in the same range. Numerous factors can be considered to contribute to the rise or fall time of individual SI events. First to be considered are the speed and thickness of the corresponding solar-wind structure. The signal observed at a given point inside the magnetosphere is the summation of all the waves that are generated successively at all the points on the magnetopause with delays reflecting the sweeping motion of the responsible solar-wind structure. Hence the rise or fall time would be correspondingly longer if the speed of the solar-wind structure is slower. A weak anticorrelation has indeed been found between the rise time of SI and the mean transmission speed of the responsible shock wave as estimated from the transit time between the solar flare and the SI (Nishida, 1966b). The thickness of the solar-wind structure would obviously influence the rise or fall time. Other factors that are considered to be important arise from the hydromagnetic wave propagation inside the magnetosphere. These include the differences in transit times of the compressional (or expansive) waves arriving at an observation site from different points on the magnetopause (Dessler et al., 1960) and the repetition of the signal due to multiple reflections of the waves between the magnetopause and the ionosphere (Willis, 1964). All these factors seem to have time scales of one to several minutes.

One of the possible effects of sudden overall compressions and expansions of the magnetosphere is the modulation of the pitch angle distribution of energetic particles trapped inside. What appears to be the consequence of such a modulation has been detected on the ground in the intensity of natural VLF electromagnetic waves and in the degree of cosmic noise absorption (CNA). An example is shown in Figure 11. There are five panels in this Figure that are labeled A through E on the left-hand side. In panel E magnetograms from representative low-latitude stations are reproduced. They demonstrate the occurrence of numerous fluctuations, whose peaks are emphasized by arrows, on a world-wide scale. These fluctuations are observed also in the tail of the magnetosphere (see panel D) with a slight delay. In view of their wide-scale coherence, these fluctuations are considered to reflect repeated compressions and expansions of the magnetosphere; namely, they can be classified as sudden impulses, although the term impulse would not be quite appropriate for describing the outlook of these events.

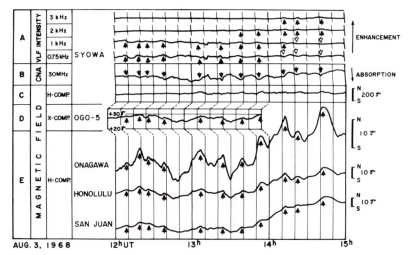

Fig. 11. Enhancement of the VLF wave intensity (panel A) and cosmic noise absorption (panel B) associated with sudden impulses (panel E). Correlated peaks are indicated by arrows (Saito et al., 1974)

Three panels at the top of Figure 11 show observations at Syowa, located at the geomagnetic latitude of $-70°$, of VLF wave intensity (panel A), cosmic noise absorption (panel B), and magnetic field (horizontal component, panel C). It is seen that both VLF intensity and cosmic noise absorption varied coherently with sudden impulses, as indicated by black arrows, in such a way that they were stronger when the magnetosphere seemed to be more compressed.

The source of the VLF waves has been considered to be the electron cyclotron instability due to pitch-angle anisotropy of the trapped particles (Kennel and Petschek, 1966). When the magnetosphere is compressed the pitch angle anisotropy is enhanced, because the kinetic energy $W_\perp$ of the motion perpendicular to the magnetic field is more enhanced than the energy $W_\parallel$ parallel to the field (Kosik, 1971). Hence an increase in the rate of the electron cyclotron wave excitation would result, and this seems to be what is observed. The theory also indicates that the central frequency of the excited wave would increase when the magnetic field becomes stronger, and this point has also been confirmed by a detailed analysis of the VLF wave observation during SI events (Hayashi et al., 1968). The excitation of electron cyclotron waves is associated with scattering in the pitch angle of parent particles. Thus more particles would flow into the loss cone and precipitate to the ionosphere when stronger waves are excited. The enhancement in the precipitating flux of energetic particles would cause enhanced ionization in the lower ionosphere where collision frequencies of electrons with neutral constituents are high, and it would thus raise the absorption of the cosmic radio noise, as observed. The correlation between daytime enhancements in cosmic noise absorption (CNA) and positive SIs and also between daytime decreases in CNA and negative SIs has been established by numerous case studies (e.g., Brown, 1973).

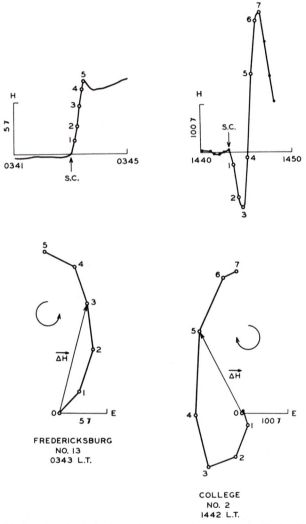

Fig. 12. Sudden impulses in the horizontal component magnetogram (top) and corresponding vector diagrams (bottom). The vector diagram is constructed by connecting the end points of the total horizontal disturbance vector as a function of time (Wilson and Sugiura, 1961)

Polarization of the SI Field

So far we have dealt with those aspects of the SI phenomenon that are
related directly to overall compressions or expansions of the magnetosphere. From the geomagnetic analysis it has long been known, however,
that an overall compression or expansion is not the sole magnetospheric
process that takes place during an SI event. It has been recognized that
the SI field observed on the ground is not merely an increase or decrease
in an almost uniform perturbation field as would be expected from the
overall compression or expansion alone. Instead, the SI field involves
complex but systematic changes in the field direction.

One of these complexities is shown in Figure 12. Two examples are
displayed together in this Figure. The variation of the magnitude of the
horizontal component is shown at the top and that of its direction is
shown at the bottom. (These SIs are designated as SCs because they
preceded magnetic storms.) It is seen that the polarization of the $\Delta \overline{H}$
vector is not linear but clearly elliptical, the sense being counterclockwise
for one case and clockwise for another. Elliptical polarizations of the SI
field like these have been found to occur often in middle and high latitudes, and the sense of polarization has been found to depend on local
time. As summarized in Figure 13, in the northern hemisphere the polarization is counterclockwise in the morning sector and clockwise in the
evening sector, the transition zone being centered at 10 LT and 22 LT.
The polarization of the SI field in the southern hemisphere is opposite
to the foregoing (Wilson and Sugiura, 1961). Note that the sense of the
polarization is expressed above by viewing the $\Delta \overline{H}$ vector downward (i.e.,
toward the earth) in each hemisphere.

The observed sense of the polarization is consistent with what is
expected for surface waves that are generated on the magnetopause and
transmitted from the nose region toward the tail. Let us take the ξ-axis
in the direction of the outward normal and η- and ζ-axes tangential to
the magnetopause. We assume that the wavelength is smaller than the
dimension ($\sim 10R_E$) of the day-side magnetosphere and we represent the
solution for the magnetic perturbation in the form of a plane surface
wave: $\boldsymbol{b} = \boldsymbol{b}_0 \exp\left[i(k_\eta \eta - \omega t) + k_\xi \xi\right]$ where $\omega > 0$. k_ξ is real and positive because the wave is assumed to be evanescent inside the magneto-
sphere where $\xi < 0$. Substituting this expression for $\boldsymbol{b}$ into div $\boldsymbol{b} = 0$,
we obtain

$$k_\xi b_{0\xi} + ik_\eta b_{0\eta} = 0,$$

namely,

$$\frac{b_{0\xi}}{b_{0\eta}} = -i\frac{k_\eta}{k_\xi}. \tag{49}$$

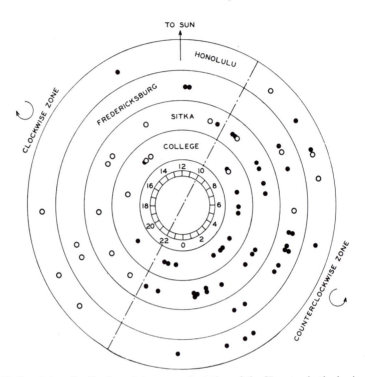

Fig. 13. Local time distribution of the sense of rotation of the SI vector in the horizontal plane as viewed from above in the northern hemisphere. (●) Counterclockwise rotation, (○) clockwise rotation (Wilson and Sugiura, 1961). Geomagnetic latitudes of the stations are: College (65°), Sitka (60°), Fredericksburg (50°), and Honolulu (21°)

This means that if $k_\eta > 0$ the sense of polarization of the **b** vector is counterclockwise, and if $k_\eta < 0$ it is clockwise, when viewed from the direction of the ζ-axis. Now let us take the ζ-axis in the direction of the magnetic field (and hence η-axis in the eastward direction) and project **b** to the ground along magnetic lines of force. Then noting that the sense of the polarization defined by viewing **b** downward in the northern hemisphere is opposite to that defined relative to the ζ-axis, we can see that the expected and the observed senses of the polarization agree if $k_\eta > 0$ after ~ 10 LT and $k_\eta < 0$ before ~ 10 LT, that is, if the surface wave is propagating tailward from the ~ 10 LT meridian.

Such a surface wave would indeed be generated on the magnetopause as illustrated in Figure 14. The solar-wind structure causing the enhancement (or reduction) in the dynamic pressure sweeps the magnetopause with a finite speed starting from the nose region. On the rear side of that structure the magnetopause moves inward (or outward). In consequence

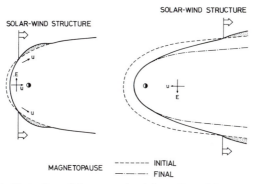

Fig. 14. Schematic illustration of the successive deformation of the magnetopause caused by the passage of a solar-wind structure leading to the reduction in the magnetospheric dimension. The region of the surface wave is hatched, and E and u show directions of electric field and motion induced inside the magnetosphere

of this inward (or outward) motion the compressive (or expansive) wave is generated in the magnetosphere. Since this wave travels along the magnetopause faster than the solar-wind structure sweeps the magnetopause, on the forward side of the solar-wind structure the magnetospheric pressure would be raised (or reduced) above (or below) the dynamic pressure of the solar wind that is locally in contact. This would cause the magnetopause to expand (or shrink) on the forward side of the structure while on the rear side it has become compressed (or expanded) due to the changed state of the solar-wind dynamic pressure. Thus the stress exerted on the magnetopause on the rear side of the solar-wind structure is released partly by the deformation of the magnetopause. The rear edge of this deformation travels along the magnetopause surface with the solar-wind structure concerned, and k_η is expected to be plus or minus in post- and prenoon hours, respectively. The reason why the transition of the polarization has been observed at ~ 10 LT rather than at ~ 12 LT is not yet fully understood, but it would be due in part to the aberration of the solar-wind velocity.

I.5 Ionospheric Modification of the SI Signal

Hydromagnetic propagation of the SI signal is accompanied by the electric pulse that is related to the magnetic signature of SI by the rot $E = -\dfrac{\partial B}{\partial t}$ relation. When the signal reaches the ionosphere this electric pulse generates a transient flow of electric current in the ionosphere, and the resulting magnetic field adds to the field transmitted directly from the outer magnetosphere.

The presence of the ionospheric contribution to SI fields observed on the ground is most evident in the equatorial region. In Figure 15a magnetograms (horizontal components) of three cases of positive SIs obtained

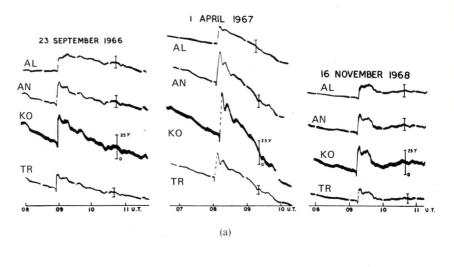

(a)

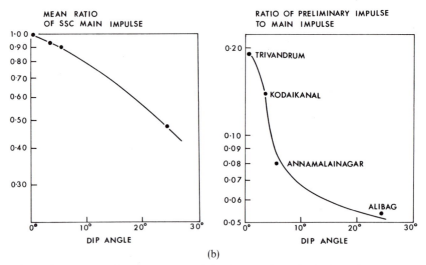

(b)

Fig. 15a,b. (a) Equatorial enhancement of positive SIs shown by horizontal component magnetograms from Indian observatories, and (b) latitudinal variation of the amplitude of the main impulse of positive SI (left) and of the ratio of preliminary impulse to main impulse of positive SI for Indian stations (right) (Rastogi and Sastri, 1974). In (a) the first two letters of the station names are used as station codes

by Indian observatories are compared. Records from four stations are displayed for each case in the order of the decreasing dip angle[4]. It is seen that the amplitude of positive SIs sharply increases toward the equator. This tendency has long been known as the 'equatorial enhancement' of SI, and the mean ratio of SI amplitude relative to the equatorial value is plotted in the left panel of Figure 15b as a function of the dip angle. The equatorial enhancement has been noted equally for the amplitude of negative SIs, as will be exemplified in Figure 16b. It is very unlikely that these enhancements occurring within a narrow strip of ~ 1000 km width are caused by currents flowing high up in the magnetosphere, and hence we are led to consider that part of the observed SI field is due to modulation of the ionospheric current.

Preliminary and Following Impulses of SI

The H-component increases of positive SIs in Figure 15a are preceded by brief decreases. The amplitude of these 'preliminary impulses' is enhanced at the equator even more significantly than the amplitude of the 'main impulse' that immediately follows: As is seen in the right panel of Figure 15b the ratio of amplitude of the preliminary impulse to that of the main impulse increases sharply toward the equator. For negative SIs the preliminary impulse takes the form of a brief increase in the horizontal component as is exemplified by the Koror, College, and Eskdalemuir records in Figure 16b. The duration of the preliminary impulse is usually less than a minute. (Positive SIs and SCs accompanied by the preliminary impulse have been designated as SI* or ⁻SI and SSC* or ⁻SC.) The occurrence frequency of the preliminary impulse tends to be high in the region covered by dots in Figure 16c. At College (65°) in the afternoon, as well as at Koror ($-3°$) in the early afternoon, its occurrence frequency is about 60 to 70%, and the occurrences at these two locations are well correlated (Araki, 1976). At Honolulu (21°), on the other hand, clear cases of preliminary impulses are rare, but a faint pulse can often be recognized in the rapid-run magnetogram of this station when clear preliminary impulses are being observed at auroral-zone and equatorial stations (Nishida and Jacobs, 1962a). (Recall that an SI event is classified as

---

[4] Since the geomagnetic field is not exactly given by the centered dipole, it is sometimes more appropriate to use the dip angle I of the geomagnetic field with respect to the horizontal plane, rather than the geomagnetic latitude θ, to express the distance of the observing site from the equator. The angle $\tan^{-1}(\frac{1}{2}\tan I)$ is called dip latitude because for the pure dipole field θ and I are related by $\tan I = 2 \tan \theta$.

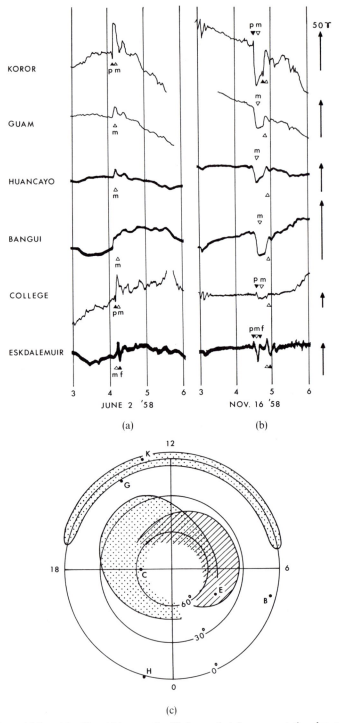

Fig. 16a–c. (a) A positive SI and (b) a negative SI observed at six representative observatories. Arrows designated as *p*, *m*, and *f* indicate preliminary, main, and following impulses. (c) Distribution of the shape of SI. Preliminary impulse tends to appear in the dotted region, and following impulse tends to appear in the hatched region. Position of six observatories whose records are reproduced in (a) and (b) are indicated for ∼5 UT (after Nishida and Jacobs, 1962a; Matsushita, 1962; Araki 1976)

positive SI or negative SI according to whether the main impulse, which appears globally, is an increase or a decrease in the H component.)

Another type of complexity in the shape of SI has been recognized in the hatched region of Figure 16c. As exemplified in Figure 16a by the records from Eskdalemuir, the main impulse of a positive SI in that region is immediately followed by an H-component decrease, and as seen in Figure 16b, the main impulse of a negative SI is followed by an H-component increase there. In these samples the amplitude of the 'following impulse' (f) is comparable to that of the main impulse (m), but this is not always the case. Nevertheless, the immediate occurrence of the following impulse tends to make SIs in that region less clearly defined and their amplitudes suppressed. The duration of following impulses is usually several to ten minutes. Positive SIs and SCs accompanied by following impulses have sometimes been referred to as SI$^-$ and SC$^-$ (Matsushita, 1962). (In the November 16 case given in Figure 16a, the negative SI discussed above is followed about 20 min later by a positive SI. Preliminary and following impulses for this event can be recognized in the records from Koror and Eskdalemuir, respectively.)

The global distribution of the SI field has been expressed by 'equivalent' current systems like those shown in Figure 17. The equivalent current system is a hypothetical flow of electric current in the ionosphere that corresponds to the magnetic perturbation field observed on the ground. Whether or not the current actually flows in the ionosphere has

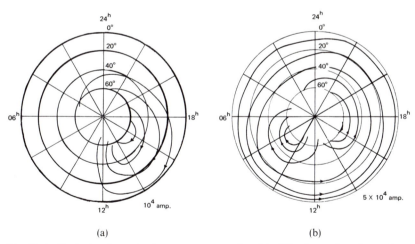

(a)   (b)

Fig. 17a,b. Equivalent ionospheric current systems of positive SI in the northern hemisphere (a) in the early stage when the preliminary impulse started and (b) in the later stage when the following impulse peaked. Numerical figures at bottom right indicate the current flowing between adjacent streamlines (Nishida and Jacobs, 1962a)

to be checked by other means, but the 'equivalent' current system serves as a convenient and useful representation of the global distribution of the perturbation field. Figure 17a corresponds to the stage when the preliminary impulse appeared in the horizontal component in a limited part of the world, while Figure 17b corresponds to a later stage when the following impulse reached its peak. It is seen that the equivalent current system for the earlier stage is basically a current vortex occupying the afternoon sector. The equivalent current vortex for the later stage is the combination of two vortices and the zonal current. The zonal part is obviously the 'equivalent' representation of the current flowing at higher levels of the magnetosphere. For negative SIs directions of the current flow are opposite to those given in Figure 17 (Nishida and Jacobs, 1962a,b).

The equivalent current systems of Figure 17 have to be taken with some reservation. In the case of Figure 17a it may not be appropriate to express the ionospheric current in the earlier stage of SI by a closed current system, since the duration of the preliminary impulse is comparable to the time taken for the disturbance to spread over the world. Moreover, it is not clear if the equatorial current is really connected to high latitude currents, in view of the apparently frequent absence of preliminary impulse in intermediate latitudes. In the case of Figure 17b representing the later stage of SI, it is not yet known precisely how much of the current is really ionospheric, because no attempt has been made to separate ionospheric and magnetospheric contributions by comparison of simultaneous ground and satellite observations.

Thus present morphology of the ionospheric contribution to SI is still less than satisfactory. Nevertheless, let us describe an available theoretical model for future comparison with space-probe data. In this model the ionospheric currents for both earlier and later stages of an SI event are considered to comprise two current vortices in each hemisphere. These current vortices are thought to be centered, in both stages, around dawn and dusk sides of the polar-cap boundary, but the direction of the current flow is supposed to reverse as the event develops from the earlier stage to the later stage. As will be discussed in Section II.2, such current vortices essentially represent ionospheric Hall currents. Then the foci of the current vortices can be taken to represent maximum or minimum of the electric potential, and the observed sense of the current vortices suggests that for positive SI the electric field across the polar cap is directed from dusk to dawn in the earlier stage and dawn to dusk in the later stage, respectively.

The theory points out that the electric fields having the assigned polarity can be produced in the magnetosphere as compression occurs progressively from the day side to the night side of the magnetosphere. In the earlier stage of a positive SI, when only the day side of the magnetosphere is experiencing the inward motion, the dusk-to-dawn electric field is produced, as illustrated in Figure 14a, by virtue of the $E = -u \times B$ relation. In the later stage, when the solar-wind structure has proceeded to the night side, the compression occurs in the magnetotail, and the dawn-to-dusk electric field would be produced in association with the earthward motion of the tail plasma as illustrated in Figure 14b. In both stages the extremes of the electric potential are on the magnetopause, and they would be projected along magnetic lines of force to the boundary of the polar cap in the ionosphere. Thus it is conceivable that the current vortices with the expected sense and geometry would be produced in the ionosphere when these electric fields successively reach that region, accompanied by the compressive motion and increase in the magnetic field (Tamao, 1964). The electric fields with polarities opposite to the foregoing would be produced during a negative SI when the magnetospheric expansion proceeds from the day side to the night side. A theory has also been proposed in which the preliminary and the following impulses of SI are attributed to the ionospheric screening effect (Nishida, 1964b). These two theories are basically identical, however, as will be discussed in Section V.5.

The morphology of the SI shapes as summarized by equivalent current systems of Figure 17 should in fact be combined with the morphology of the polarization of the SI vector shown in Figure 13. Correspondingly, the successive generation of the compressive (or, expansive) wave in different regions of the magnetosphere should be studied simultaneously with the propagation of the surface wave on the magnetopause. A complex initial and boundary value problem remains to be solved for a complete understanding of the SI phenomenon.

The mechanism by which the ionospheric current flowing zonally along the dayside dip equator is enhanced and the equatorial enhancement is produced in the amplitude of geomagnetic variations will be discussed in Section IV.5. For the time being let us note that the greater enhancement of the preliminary impulse as compared with the main impulse probably reflects the greater relative contribution of the ionospheric current to the former than to the latter; while the preliminary impulse is due (probably) entirely to the ionospheric current, a substantial part of the main impulse should be produced by the magnetospheric current.

II. Power Supply Through the Interplanetary Field Effect

II.1 Introduction: The Reconnected Magnetosphere

Mounting evidence has led us to believe that the transport of the solar-wind energy across the magnetopause occurs principally by the 'reconnection' mechanism. According to this mechanism the penetration of the interplanetary electric field into the magnetosphere is expected to accompany the inflow of the solar wind energy. Geomagnetic disturbances resulting from the ionospheric current driven by the penetrating electric field have been identified as 'DP2' and 'DPY' (or, 'Svalgaard-Mansurov effect').

Because the solar wind is the hot plasma having high electric conductivity, the electric field in it should be practically zero when viewed in the frame of reference moving with the solar wind. This means that in the framework of the earth's magnetosphere, which is moving with velocity $-v_s$ relative to the solar-wind frame, the electric field E_I exists, which is given by

$$E_I = -v_s \times B_I \tag{50}$$

where B_I is the interplanetary magnetic field. The magnitude of the interplanetary electric field E_I is appreciably large; with $v_s = 300$ km/s and $B_I = 5\gamma$, the electric potential difference arising from E_I over the lateral dimension $\sim 40 R_E$ of the magnetosphere is 300 kV, which is sufficiently large as compared with the electric potential that is considered to exist inside the magnetosphere.

The idea that the input of the solar-wind energy into the magnetosphere occurs by way of the penetration of E_I into the magnetosphere was originally suggested by Alfven (1939). Before this idea became generally accepted, however, one serious difficulty had to be overcome. As explained in the last Chapter the magnetopause surface derived from the pressure balance condition alone is covered entirely by magnetic field lines that meet together at two singular points on the surface: northern and southern neutral points. This means that an arbitrary pair of points on the magnetopause is connected by field lines via neutral points, and it should have equal electric potential if the electric conductivity along magnetic field lines is as high as given by the ordinary Spitzer formula ($\sigma_0 = Ne^2/mv$). Namely, the magnetopause represents a surface acting as an electric shield and the electric field E_1 of interplanetary origin cannot penetrate into the magnetosphere.

If, however, magnetic field lines inside the magnetosphere are directly connected with those in the interplanetary space, the electric field can be communicated via connected field lines. Dungey (1962) proposed just such a model for the magnetosphere, which is now known as the 'open' model of the magnetosphere. The topology of magnetic field lines in the open magnetosphere is illustrated schematically in Figure 18. Since the penetration of the electric field is associated with the generation of the bulk motion inside the magnetosphere, the plasma, as well as the field lines frozen in it, is convected in the open magnetosphere in the direction indicated by arrows in this Figure. At the day-side reconnection line interplanetary and geomagnetic field lines approach each other and are reconnected, and they form a pair of 'open' field lines that extends from the earth to the interplanetary space. The region encompassing the ground projection of the field lines of this kind will be referred to as the polar cap. At the night-side reconnection line the preceding pair of the open field lines makes contact and is reconnected to yield a geomagnetic closed field line (having both ends on the earth) and an interplanetary field line. An essential assumption underlying the open magnetosphere model is that the electric conductivity can be self-consistently reduced at these reconnection lines, so that the frozen-in field theorem is violated and the field-line reconnection is achieved.

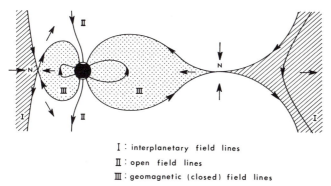

I : interplanetary field lines
II : open field lines
III : geomagnetic (closed) field lines

Fig. 18. Topology of the magnetic field and direction of the plasma flow in the noon–midnight section of the open magnetosphere (Dungey, 1962). When the IMF direction is due southward the neutral points N represent sections of the neutral ring that surrounds the earth

Since geomagnetic field lines from the polar cap extend to the solar wind, in the open magnetosphere the magnetopause does not represent a tangential discontinuity as it does in the closed magnetosphere model discussed in the last Chapter. The magnetopause of the open magnetosphere is expected to be a complex structure that consists of a set of slow and intermediate shock waves (Petschek, 1966). Verification of the openness of the magnetosphere in terms of the magnetopause structure, however, is not yet fully accomplished. Attempts to do so have encountered great difficulties because local attitudes and positions of the magnetopause keep changing, probably due to surface waves excited on the magnetopause. Nevertheless, the open magnetosphere model has become widely accepted because correlations observed between solar-wind parameters and magnetospheric behaviors are basically consistent with what is expected from this model.

This Chapter is concerned with geomagnetic observations that are considered to represent direct consequences of the day-side reconnection process. Since the magnetospheric electric field depends on E_I and hence on B_I, the correlation with the interplanetary magnetic field is the characteristic signature of geomagnetic disturbances of this kind. We shall deal successively with disturbance fields that are correlated with southward, east–west, and northward components of B_I, and discuss the physics of the open magnetosphere on the basis of these and related observations in the magnetosphere. The solar magnetospheric coordinate system, which refers to directions both of the solar wind and of the geomagnetic dipole, provides a suitable framework for the topic of this Chapter. The 'interplanetary magnetic field' will subsequently be abbreviated as IMF.

II.2 Geomagnetic Response to the Southward Component of IMF

The electric field and the convection velocity are related by the frozen-in condition $E = -u \times B$. This relation is applicable in the magnetosphere where the collision frequency is much less than the gyrofrequency both for ions and electrons. At altitudes below about 160 km, however, the ion collision frequency is higher than the gyrofrequency and the convective motion of ions is impeded due to frequent impacts with neutral particles. Similar altitude limit for the electron convection lies around 90 km. In the intermediate height range of 160 ∼ 90 km where only electrons participate in the convective motion, an electric current is produced. This current, known as the ionospheric Hall current, flows in the direction opposite to the direction of the convection velocity (Axford and Hines, 1961). In addition, the Pedersen current, which flows in the direction of the electric field, is also produced in the ionosphere.

Thus when there are electric field and convection in the magnetosphere, electric currents flow in the ionosphere. Geomagnetic effects of these currents should be detectable on the ground, and, if the reconnection is indeed the source of the electric field, they should show dependence on IMF conditions. This section deals with geomagnetic variations which are dependent on the southward component of IMF.

DP2 Fluctuations

The geomagnetic effect of the interplanetary magnetic field can be easily detected when the ground magnetic field is found to simulate the time variation of IMF. Figure 19 shows an example of the geomagnetic variation that has such a character. The variations concerned are the quasi-periodic fluctuations, called DP2, which are seen in Figure 19b. Characteristically, these fluctuations with a quasi-period of around 1 h are

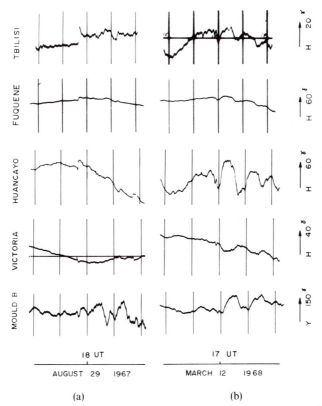

Fig. 19a,b. An example of DP2 (b) as compared to that of SI (onset time: ~1740 UT) (a). (Nishida and Maezawa, 1971)

recorded coherently all over the world from the pole [e.g., Mould Bay (geomag. lat. 80°)] to the equator [e.g., Huancayo ($-1°$)]. The distur-bance of this mode is strongly enhanced at the dayside equator as can be seen by comparing the Fuquene (17°) and Huancayo ($-1°$) records in Figure 19b; in the polar cap its magnitude varies significantly with the seasons, being stronger in summer than in winter (Nishida, 1968). These features suggest that the DP2 disturbance field is due mainly to currents flowing in the ionosphere. The comparison with the record of SI shown in Figure 19a reveals the following distinction between DP2 and SI; in middle latitudes on the morning side [i.e., at Victoria (54°)], an SI is initiated by the main impulse, which, in the H component, has the same sign as the SI field observed everywhere in middle and low latitudes. Namely, for a positive SI an increase in the H component is observed at Victoria, though briefly, as at other stations in middle and low latitudes. For DP2, on the other hand, the phase of the H-component variation at Victoria is opposite to that of H observed at other middle and low lati-tude stations (at Huancayo for example) throughout the event.

The correlation of the foregoing DP2 event with the southward com-ponent of IMF is demonstrated in Figure 20. In the left panel the tracing of the H-component record at Huancayo is compared with simultaneous observations of IMFs north–south component and solar-wind dynamic

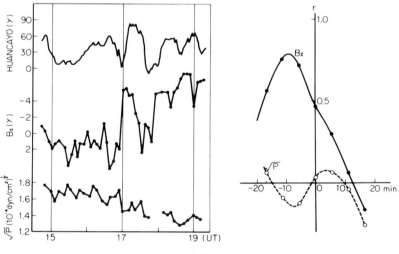

MARCH 12, 1968

Fig. 20. Comparison of DP2 fluctuations with variations in B_z and $\sqrt{P}$ in the solar wind. *Left panel: visual comparison of records; right panel: cross-correlation analysis.* Negative time lag means that the interplanetary observations precede the ground observation (Nishida and Maezawa, 1971)

pressure. DP2 fluctuations correspond peak-to-peak to B_z but are poorly correlated with $\sqrt{P}$. This impression is confirmed by the cross-correlation analysis (right panel), which shows that the H component at Huancayo during this 4-h interval has a correlation coefficient of 0.8 with B_z observed about 10 min earlier while it is essentially independent of $\sqrt{P}$. [Prior to the cross-correlational analysis, a low-cut filter was applied to the data to eliminate the influence of the slowly varying component (Nishida and Maezawa, 1971).]

A similar comparison between interplanetary and geomagnetic observations is made in Figure 21. Here the top panel contains records of density n, velocity v, IMF strength B, and IMF latitudinal angle θ in the solar magnetospheric coordinates. Since both n and v varied little, the dynamic pressure was kept nearly constant during the interval, and since

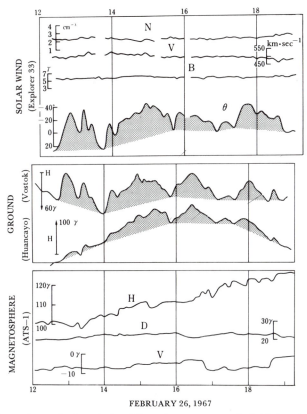

FEBRUARY 26, 1967

Fig. 21. Comparison of solar wind and interplanetary magnetic field data (top), ground magnetograms (middle), and magnetic field data in the inner magnetosphere (bottom). DP2 fluctuations and associated IMF variations are hatched (Nishida, 1971b)

B was almost fixed, the variation in B_z resulted mainly from the variation in θ. The middle panel contains tracings of magnetograms from Vostok ($-89°$) located at the southern geomagnetic pole and from the equatorial station Huancayo ($-1°$). Clear peak-to-peak correlation is noted between the θ angle of IMF and the ground magnetic variation classified as DP2 fluctuations by the global feature. [In order to assist the comparison, hatchings are given to peaks (or troughs) in θ and DP2 variations.] The last panel contains magnetic field data obtained inside the magnetosphere at the radial distance of $6.6R_E$, and this record demonstrates that the field strength in the inner magnetosphere did not coherently respond to the IMF θ (or B_z) variation. Thus the interaction process connecting B_z and DP2 is distinct from the modulation of the magnetospheric dimension by changes in the dynamic pressure of the solar wind.

Since the solar-wind structures producing DP2 fluctuations like the preceding ones are characterized by significant fluctuations in θ (or B_z) that are uncorrelated with *n*, *v*, or *B*, they belong to either of the following two categories: (1) a subset of tangential discontinuities having little variation in *p* and *B*, or (2) intermediate shock waves. (Here the term Alfven wave would be more appropriate than intermediate shocks, since relatively gradual variations whose scale lengths are greater than the thickness of the fast shock waves are meant. The characteristics of inter-mediate shocks or Alfven waves are known to be independent of the thickness and amplitude.) In more general circumstances where solar-wind structures having variations both in nv^2 and B_z strike the magne-tosphere, geomagnetic variations of DP2 and SI modes would occur superposed.

The global distribution of the DP2 field is shown by representative magnetograms in Figure 22 and summarized by equivalent current sys-tems in Figure 23. For the reason described later the hatched peaks in Figure 22 are scaled to be the DP2 field. The dashed curve is the record on the nearby quiet day. The characteristic feature is that on the evening side, the *H* (or, *X*) component shows maxima simultaneously at auroral (Kiruna) and middle (San Fernando) latitudes, so that the equivalent current is directed from the noon to the night side through the dusk sector at both of these latitude zones. Also, at a middle-latitude station in the prenoon sector (Fredericksburg), the corresponding peaks appear as minima in *H* but maxima in *D*, so that the equivalent current flows from the noon side of high latitude to the morning side of low latitude. Consequently the equivalent current system of DP2 in each (northern or southern) hemisphere involves two large vortices that originate from the polar cap. Of these, the evening-side vortex occupies the larger area and extends well to low latitudes. The morning-side vortex, on the other hand,

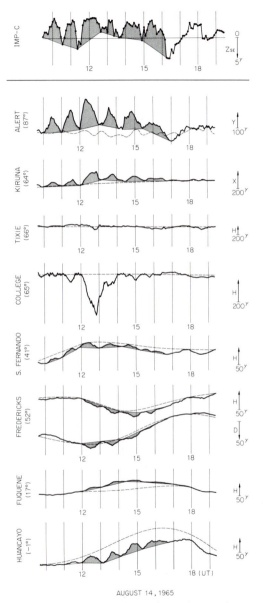

AUGUST 14, 1965

Fig. 22. A typical example of a train of DP2 fluctuations. The IMF data are given at the top for comparison, and the dashed curves represent the quiet-day record on August 28, 1975 (Nishida, 1971a)

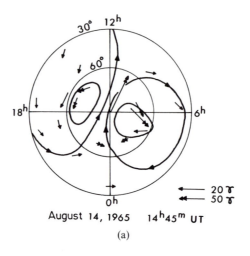

August 14, 1965   14ʰ45ᵐ UT

(a)

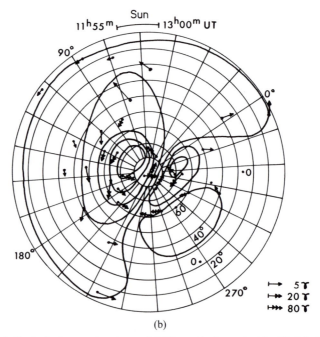

(b)

Fig. 23a–d. Equivalent current systems of DP2 events shown in Fig. 22 (Afonina et al., 1975)

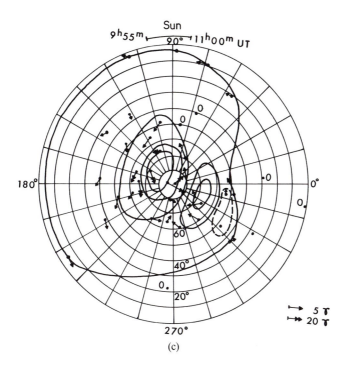

(c)

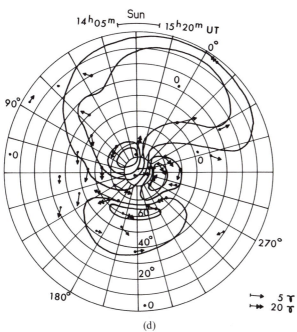

(d)

Fig. 23 (*cont.*)

is drawn also to extend to low latitudes in the equivalent current of Figure 23a (Nishida, 1971a) but is shown to terminate at auroral latitudes in Figure 23b–d (Afonina et al., 1975). The difference originates practically from the smallness of the amplitude of the DP2 field in the night-side low-latitude region. In that region the peaks that are in synchrony with the peaks observed elsewhere are usually hard to identify, and the possibility is high that disturbance fields of different modes are scaled inadvertently. Hence in the author's view the third vortex in the night-side low-latitude region included in the equivalent current systems of Figure 23b–d may reflect the coexistence of disturbance fields other than the DP2 field. [Due to the same difficulty the westward zonal current that was depicted in the equatorial region in earlier DP2 current systems (e.g., in Nishida, 1968) has to be considered indeterminate.]

The polarity of the DP2 field, or equivalently the direction of flow of the DP2 equivalent current, has been a matter of controversy. The problem has arisen as DP2 fluctuations have been detected as time-varying fields that are generally superposed on other mode(s) of geomagnetic disturbance. If DP2 were the sole mode of the disturbance field existent at a given moment, or, if DP2 were by far the largest disturbance field everywhere on the ground, it would have been possible to determine the polarity of the DP2 field by measuring the disturbance from the quiet-day level (Sq). This, unfortunately, is not true: in the case presented in Figure 22 where magnetograms showing the DP2 activity are given superposed on magnetograms recorded on a quiet day, the hatched peaks evidently represent departures from the quiet level at the polar station Alert and the eveningside auroral-zone station Kiruna; however, at the dayside equatorial station Huancayo, the simultaneous peaks are located appreciably below the quiet-day level, and it appears that field depressions between the hatched peaks represent the intrinsic disturbance field. Since it is highly unlikely that a disturbance field with a time scale of ~ 1 h has a phase difference of as much as ~ 30 min between the polar region and the equator, one is forced to conclude that either at the equator or in the polar cap (or, more plausibly, everywhere) DP2 fluctuations occurred superposed on some other disturbance field whose spatial distribution was different from that of DP2.

In circumstances where more than one mode of disturbance occurs simultaneously, filtration is one of the possible methods for separating a specific component. The scaling of the amplitude of the globally coherent quasi-periodic oscillations has been adopted to define the DP2 field exactly for this reason. This implies that the direction of the DP2 current flow has to be determined using other sources of information that allow the separation of DP2 from the other, superposed disturbance fields. The direction of the current in Figure 23 was chosen to correspond to the hatched peaks of Figure 22, which represent almost the entire disturbance field in the polar cap, because there was an indication that the H component at Huancayo has been depressed due to the superposition of a long-term disturbance field (Nishida, 1971a).

This choice of the flow direction means that when the southward component of IMF increases the disturbance current directed toward the sun is made more intense over the polar cap. The time lag between southward peaks of B_z variations and corresponding peaks of DP2 fluctuations is due primarily to the separation between the satellite monitoring the IMF and the magnetosphere. However, when the contribution from this effect is subtracted, an average residual of 15 min has been found, which is appreciably longer than the transit time of hydromagnetic waves from the magnetopause to the ground (Nishida and Maezawa, 1971).

The preceding choice was questioned by Matsushita and Balsley (1972) on the grounds that during an interval involving the DP2 (like) events the equatorial electrojet was weakened instead of being intensified, according to direct measurements using the backscatter method. However, it has been shown that the disturbance observed during the interval they examined is not due to the occurrence of a single mode of the disturbance field (Nishida, 1973), and it cannot be determined whether the weakening of the equatorial electrojet was due to DP2 or to other, superposed, modes of disturbances. Unless the nature of that superposed disturbance is clarified and its spatial distribution understood, their observation of the electrojet weakening cannot serve as decisive information for deducing the direction of the DP2 current flow.

As long as DP2 is separated from the background by noting the occurrence of coherent, world-wide field fluctuations, the identification of DP2 has to be limited to those occasions where the field fluctuations of the DP2 mode stand out from the background as well-defined wave trains; if DP2 appears as a constant field component or as a gradually increasing or decreasing trend, it is not possible to identify them by the preceding method. Hence case studies of the B_z-DP2 relationship have been restricted to the cases where B_z of IMF is oscillating. In order to establish the geomagnetic response to B_z for more general circumstances, it is necessary to complement the foregoing study with analyses using statistical methods.

Thus Maezawa (1976) compared hourly values of both X (northward) and Y (eastward) components of the geomagnetic field at four polar-cap stations [Thule (89°), Resolute Bay (83°), Mould Bay (80°), and Godhavn

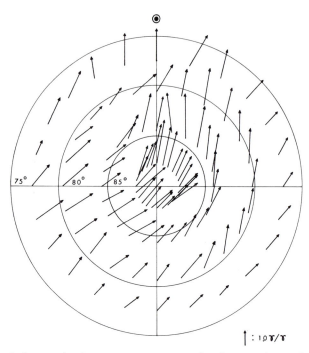

Fig. 24. Equivalent overhead current vectors representing the regression coefficient of the polar-cap magnetic field on B_z of IMF for the case of $B_z < 1\gamma$. The unit vector at the bottom right corresponds to a 10γ increase in the horizontal geomagnetic field produced by a 1γ increase in the IMF southward component (Maezawa, 1976)

(80°)] with IMFs B_y and B_z components averaged over 1-h intervals that preceded the corresponding ground observations by 30 min. To examine the dependence on the southward IMF only those cases where the hourly average of B_z was less than $+1\gamma$ were chosen, and the linear regression coefficients of X and Y on B_z were obtained for each hour in local time after eliminating the dependence on B_y. The results obtained for five summer months in 1966 and 1968 are displayed in Figure 24 in the form of the equivalent current system. (The corrected geomagnetic latitude[5] is used in this Figure.) It shows that in the polar cap the increase in the southward component of IMF is reflected as the proportional intensifica-

---

[5] The corrected geomagnetic coordinates of a given point A on the ground are calculated as follows: (1) trace the field line of the real main field of the earth from A to the equatorial plane; (2) from there return to the point B on the ground following the field line of the dipole component of the geomagnetic field; and (3) assign the geomagnetic coordinates of the point B to the corrected geomagnetic coordinates of the point A. Tables and maps of the corrected geomagnetic coordinates are given in Hakura (1965).

tion of the equivalent current directed toward the sun. This is essentially what was noted in the polar-cap portion of Figure 23, and thus the results of the statistical analysis are consistent with those obtained by case studies. [The skewing of the equivalent current in the night sector of Figure 24 has been interpreted by Maezawa (1976) to be due to the contribution of the disturbance originating from the field-aligned current connected to the magnetotail, on the basis of the examination of the time delay between B_z and corresponding geomagnetic effects.]

Electric Field Induced by Southward IMF

The equivalent current system of Figure 23 or Figure 24 represents a combined effect of the ionospheric (Hall and Pedersen) current and the current flowing in and out of the ionosphere along magnetic field lines. Hence it is not an easy task to derive the electric field from the observation of the equivalent current system, and it is more practical to compare the observed equivalent current with the one calculated on the basis of simple but physically realistic electric field models. Lyatsky et al. (1974) made such a calculation for the electric field that is produced originally by two point charges that are placed on the edge of the polar cap symmetrically with respect to the noon–midnight meridian. Charges placed in dawn and dusk sectors are assumed to be positive and negative, respectively, so that the electric field across the polar cap is originally directed from dawn side to dusk side. The ionospheric Pedersen current between these charges is assumed to be connected to (and maintained by) the field-aligned currents flowing in and out of the ionosphere, which are modeled by two line currents that extend radially from the position of these charges. The ionospheric conductivity is considered to be uniform in both day and night hemispheres, but the difference between day and night conductivities is taken into account. Charges accumulated at the conductivity discontinuity are considered to grow until the resulting secondary electric field prevents the further accumulation. The Pedersen and Hall conductivities are assumed to be equal.

The results of their calculation are reproduced in Figure 25. Figures 25a–c refer to cases where charges are placed exactly on the dawn and dusk meridians. Figure 25a represents the case where day and night conductivities are identical. The equivalent current for this case consists exactly of two symmetrical current vortices representing the Hall current; this is because the fields produced on the ground by the Pedersen current and by the vertical currents exactly cancel when the conductivity is uniform, as pointed out by Fukushima (1969). Figure 25b is for the case

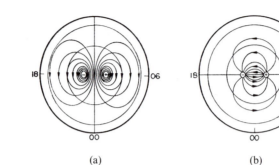

(a)   (b)

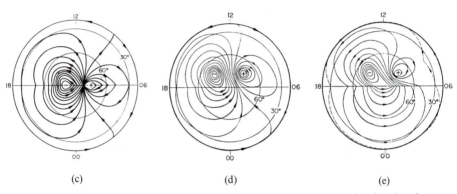

(c)   (d)   (e)

Fig. 25a–e. Equivalent current systems produced by two point charges placed at the edge of the polar cap. For details see text (Lyatsky et al., 1974)

where the night conductivity is zero, and Figure 25c, for the case where the day conductivity is twice as large as the night conductivity. Notably the day–night asymmetry in the conductivity produces an asymmetry in the size of the current vortices, and the area covered by the evening vortex increases as the day-side conductivity is increased. Figure 25d represents the case when the charges are placed in the day sector under the same conductivity condition as for Figure 25c, and it is transformed to Figure 25e when the night-side conductivity is reduced to zero. These current patterns, particularly Figures 25c and d, bear striking similarity to the observed equivalent current systems of DP2. Although there would be a difference in the night-side low-latitude region if the third vortex depicted in Figures 23c and d should turn out to be the intrinsic part of the DP2 current system, it is then conceivable that the current pattern has been modified by the formation of the high conductivity strip along the auroral oval.

The correlation observed between B_z and DP2 indicates then that the dawn-to-dusk electric field over the polar cap varies proportionally to the southward component of IMF. This inference is supported by the comparison of direct measurements of the electric field over the polar cap with simultaneous IMF observations. In Figure 26 hourly average values of the dawn-to-dusk component of the polar-cap electric field observed by balloons are compared with hourly averages of B_z for the same hour. Figures 26a and b correspond to electric-field observations over the polar-cap stations Thule (89°) and Resolute Bay (83°), and in both cases the dawn-to-dusk electric field is seen to increase as the IMF southward component increases; the least square fit is given by $E \, (\text{mV/m}) = 22 - 3B_z \, (\gamma)$ (Mozer et al., 1974).

Statistical analyses of the geomagnetic response to the southward component of IMF have also been carried out by using indices representing the magnetic activity in the auroral zone. In these cases the interpretation of the result in terms of the B_z-magnetospheric electric field relation cannot be straightforward, because auroral-zone magnetic activities are strongly influenced by conductivity enhancements due to precipitating auroral particles (Brekke et al., 1974) and because the center of the magnetic activity is sometimes located far poleward of the auroral-zone stations that contribute to available indices (Akasofu et al., 1973b). Nevertheless, the auroral-zone magnetic indices provide useful measures of the amount of energy that has been transferred to the magnetosphere from the solar wind, since the magnetospheric substorm which these

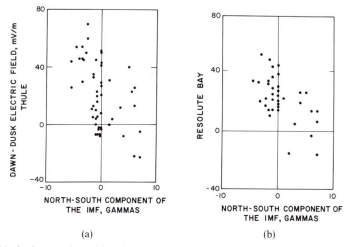

Fig. 26a–b. Scatter plots of hourly average values of the dawn-to-dusk component of the ionospheric electric field and the north–south component of IMF measured during the same hour (Mozer et al., 1974)

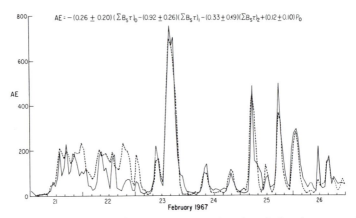

Fig. 27. A linear fit of the interplanetary parameters (· · · ·) to the hourly average of AE
(———). The subscripts on the parameters give the lead time in hours (Arnoldy, 1971)

indices represent is the major sink of that energy, and their comparison
with IMF parameters have yielded interesting results after successful
attempts were made by Fairfield and Cahill (1966) and Rostoker and
Fälthammar (1967). Figure 27 illustrates a result of such a correlational
analysis in which the hourly AE index (see footnote on p. 00 for definition)
is used. A running fit to AE (solid curve) is sought by a linear combination of IMF and solar wind parameters (dashed curve), and $(\Sigma B_s \tau)_1$,
which is the summation of the southward-directed, solar magnetospheric
B_z component of IMF averaged over the previous 1-h interval, is found
to contribute most to the formula (Arnoldy, 1971). This point will be
pursued in the next Chapter in connection with the substorm issue.

II.3 Polar-Cap Geomagnetic Response to the East–West Component of IMF

While the north–south component of IMF fluctuates around the mean
value of zero, its east–west component frequently shows the same sign
for intervals of several days or longer. This is because the large-scale
structure of IMF is the Archimedean spiral originating from the sun.
Field lines emerging from a photospheric region where the magnetic field
is directed outward develop an eastward component (when viewed toward
the sun) as they spiral, and those from a region of inward field develop a
westward component. This structure is known as the interplanetary sector
structure, and in typical circumstances the interplanetary space is divided
into two or four sectors in each of which the outward–eastward or the
inward–westward polarity prevails (Wilcox and Ness, 1965). The geomagnetic effect of this structure is the subject of this section.

DPY, or Svalgaard-Mansurov Effect

Because the sign of the east–west component B_y remains the same for intervals longer than a day, the effect of this component can be recognized even in some daily measures of the geomagnetic field (Svalgaard, 1968; Mansurov, 1969). Figures 28a and b show its effect on the shape and amplitude of daily variations in Z (downward) and X (northward) components of a polar station Resolute Bay (83°) during 1965. Open circles represent the mean daily variation found on days when the earth is within the 'away' sector (characterized by outward IMF) while the filled circles refer to the variation within the 'toward' sector (characterized by inward IMF). (The upper panel shows the original daily variation, while the lower panel shows the difference from the all-day average indicated by a dashed curve in the upper panel.) In both components the dependence on the IMF polarity is evident particularly during 1200 to 2400 UT, which is centered around the local noon, and the observed X and Z variations suggest that a nearly zonal current is generated in the daytime to the south of Resolute Bay. The direction of the zonal current depends on the sector polarity and is eastward (westward) around the northern magnetic pole

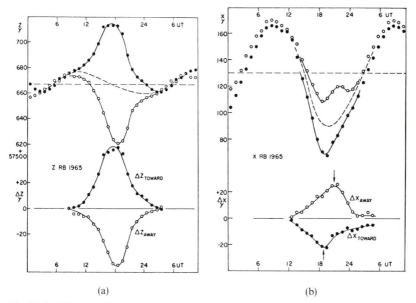

(a)   (b)

Fig. 28a,b. Effect of the IMF east–west component on the magnetic daily variation in (a) the Z component and (b) the X component at Resolute Bay. (○, ●) Variations on days when the earth is in the away and toward IMF sectors, respectively. The dashed curve in the upper panel shows an average for all days (Svalgaard, 1973)

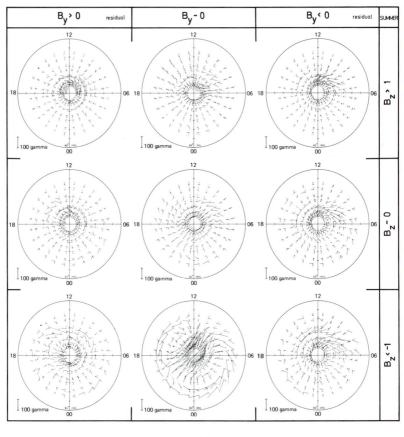

(a)

Fig. 29a,b. Matrix diagrams summarizing the average distribution of the equivalent overhead current vectors under various conditions of the B_y and B_z components of IMF: (a) in
summer and (b) in winter. The 'residual' is the difference between the field observed in the
given range of B_y and that observed when $B_y \sim 0$. (Friis-Christensen and Wilhjelm, 1975)

in the away (toward) IMF sector. Observations at other stations are
consistent with this interpretation (Svalgaard, 1973), and the feature has
sometimes been referred to as Svalgaard-Mansurov effect.

Since polarities of IMF in/out and east/west components are correlated due to the spiral structure, a question had been posed regarding
whether the in/out (B_x) or east/west (B_y) component of IMF is the real
cause of the foregoing effect. This question was examined by finding exceptional occasions where the local IMF condition deviates significantly
from a spiral so that the field is directed outward–westward or inward–

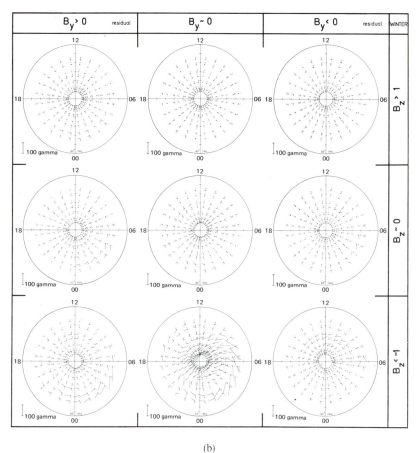

(b)

Fig. 29 (*cont.*)

eastward, and it was established that the east–west component is the real cause (Friis-Christensen et al., 1972).

A comprehensive summary of the B_y effect (DPY) observed in the high-latitude magnetic field in the northern hemisphere is shown in Figures 29a and b in the form of the equivalent overhead current distributions. In constructing these figures Friis-Christensen and Wilhjelm (1975) used hourly values of the geomagnetic data at seven high-latitude stations [Thule (89°), Resolute Bay (83°), Mould Bay (80°), Godhavn (80°), Baker Lake (74°), Fort Churchill (69°), and Great Whale River (67°)], and they divided the data into nine groups according to $\bar{B}_y$ and $\bar{B}_z$, which are 2-h averages of B_y and B_z during the same and the preceding hour. (Observations during highly variable IMF conditions were excluded.) Ranges of $\bar{B}_y$ bins are: $>0\gamma$, -1.5 to 1.5γ, $<0\gamma$, and those of $\bar{B}_z$ bins are: $< -1\gamma$, -1 to 1γ, $>1\gamma$, respectively. Summer, equinox, and winter months were treated separately, and averages were obtained in each $(\bar{B}_y, \bar{B}_z)$ category for each season. No attempt was made to remove the quiet-time daily variation. In the next step, in order to extract the $\bar{B}_y$ dependence, they subtracted the average for the cases of $\bar{B}_y \sim 0$ (namely, $-1.5\gamma < \bar{B}_y < 1.5\gamma$) shown in the central column from the averages for the cases of $\bar{B}_y > 0\gamma$ (eastward IMF) and for $\bar{B}_y < 0\gamma$ (westward IMF), and the resulting residuals are displayed in left- and right-hand columns. Figure 29a is for the summer and Figure 29b is for the winter months, respectively. The invariant latitude[6] is used here.

First let us look briefly at the central column representing the B_z effect. When $\bar{B}_z < -1\gamma$, the two-vortex system obtained in the last section by other methods is evident in the summer result, but in winter the direction of the polar-cap current flow is shifted toward dawn. This is probably because the relative importance of the perturbation related to the field-aligned current from the magnetotail rises as the intensity of the ionospheric current falls in winter due to the reduction in the high latitude conductivity. It is also noted that for the northward B_z exceeding $+1\gamma$, the current system in summer deviates significantly from DP2. These points will be examined in a later section of this Chapter.

The residual currents representing DPY have a spiral form whose sense of rotation around the northern magnetic pole is eastward for the eastward IMF but is westward for the westward IMF. Essentially the same residual current patterns are obtained for all ranges of $\bar{B}_z$, thus demonstrating that B_y- and B_z-effects are independent phenomena. The spatial extent of the residual current system, however, seems to be governed by B_z, and the area covered by the residual current system shrinks roughly by $4°$ in latitudes as $\bar{B}_z$ increases from -3 to 3γ. The intensity of the residual current is significantly reduced in winter, and in summer the current is weaker on the night side than on the day side. Thus, the current intensity is apparently controlled by the ionospheric conductivity (Friis-Christensen and Wilhjelm, 1975). That DPY observed in the ground magnetic field is indeed due to the ionospheric current has been confirmed by satellite observations of the magnetic field above the ionosphere

---

[6] The invariant latitude Λ is derived from the magnetic shell parameter L of the field line passing the given point by $\Lambda = \cos^{-1} (1/L)^{1/2}$ (McIlwain, 1961).

(Langel, 1974). For a given latitude, local time, season, and B_z, the intensity of DPY has been found to be proportional to B_y.

The response time of the polar magnetic field to the IMF east–west component is studied in Figure 30 by a comparison of records having a finer time resolution. The Z component record at Thule (89°) (solid curve) is shown superposed on the B_y component of IMF (dashed curve) shifted visually so as to reveal the best correlation. It is seen that in the sun-lit hours (1000–2200 UT) at Thule there is a clear correlation between the

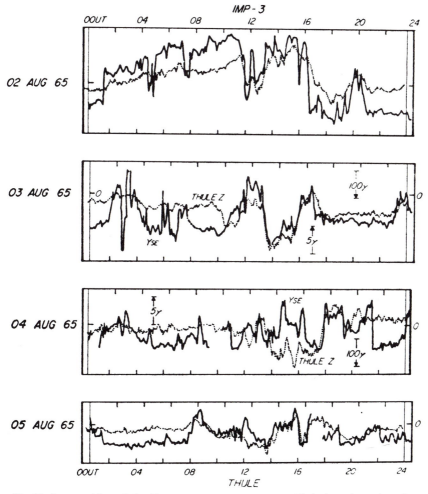

Fig. 30. Superposition of the Z-component magnetogram at Thule (-----) on the solar ecliptic eastward component Y_{SE} of IMF (———) (Kawasaki et al., 1973)

two records, and the delay time of about 25 min has been noted between them (Kawasaki et al., 1973). When the delay due to the transit between the satellite and the magnetopause is corrected, the response time of the magnetosphere to the B_y component of IMF seems to be similar to the response time recognized for the southward IMF component.

The observation that the east–west IMF component marks a well-defined signature on the polar magnetic field has promoted the idea of inferring the IMF polarity from the polar magnetic data. The idea attracted much interest because it offers the possibility of tracking the solar magnetic structure prior to the advent of the IMF observations by space vehicles. Data from the Godhavn observatory (80°), which has been in operation since 1926, are regarded as the most valuable resource. Figure 31 shows a series of the reduced H data of Godhavn for the month of May 1968. In this presentation, the monthly mean value has been subtracted from hourly mean values during intervals of 15–22 UT but the level is set to zero outside these intervals. + and − signs in Figure 31 represent IMF polarities actually observed by spacecrafts, and it is seen that a positive spike is associated with away polarity (denoted by a + sign) and a negative spike with a toward polarity (−) on most days, although there are some exceptions. The occurrence of the exceptional cases can be attributed to at least three factors: (1) data during only seven hours a day, in which the B_y effect stands out at Godhavn, are used; (2) the IMF structure sometimes deviates from the spiral, and the eastward (westward) B_y is not always associated with the outward (inward) B_x; (3) the background field on which the B_y-dependent field is superposed

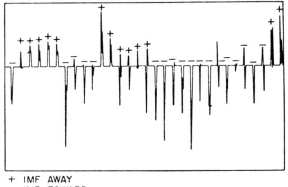

+ IMF AWAY
− IMF TOWARD

Fig. 31. Attempt to infer the IMF polarity from the geomagnetic data: Godhavn data for the month of May, 1968 are reduced in the way described in the text and are compared with the IMF polarity actually observed (Svalgaard, 1975)

is not at all constant, and sometimes its diurnal variation can be confused with the B_y-dependent signature (Svalgaard, 1975). The method can be improved when the spatial distribution of the disturbance field is examined through the use of data from other stations and when efforts are made to discriminate disturbances belonging to other modes.

Recently a system has been established to infer the polarity of IMF in near real time on the basis of geomagnetic observations at Vostok and Thule. The result is published monthly in the *Solar Geophysical Data* from NOAA. During 1972 the agreement between the observed and the inferred polarity was 87% (Wilcox et al., 1975).

Returning to Figure 29, we can see that the superposition of the residual $\bar{B}_y \gtrsim 0$ current on the two-vortex current system for $\bar{B}_y \sim 0$ and $\bar{B}_z \lesssim 0$ has the effect of creating asymmetry in the sizes of the two vortices. In the northern hemisphere, when B_y is eastward ($B_y > 0$), the evening vortex is made larger than the morning one, and when B_y is westward the morning vortex is made larger. If the equivalent current is due mainly (though obviously not entirely) to the ionospheric Hall current, the corresponding asymmetry is expected to exist in the electric potential distribution. Figure 32 shows hourly average electric field vectors obtained by three balloon flights in the northern polar cap. Solid curves in these Figures are equipotential contours drawn on the assumption that the electric equipotentials can be represented basically by a two-vortex pattern, and the expected asymmetry is indeed recognized in the sizes of potential vortices (Mozer et al., 1974). It is also seen that the intensity of the dawn-to-dusk electric field is stronger on the dawn side of the pole when B_y is eastward (Fig. 32a) and on the dusk side when B_y is westward (Fig. 32b). Consistent results have also been obtained by satellite observations of the electric field (Heppner, 1972). The asymmetry of the potential vortices can be modeled by introducing asymmetry (with respect to the noon–midnight meridian) in distributions of the positive and negative charges placed on morning and evening sides of the polar cap.

In the southern polar cap the effect of the east–west component of IMF has been found to appear with a distinctly opposite phase relative to that observed in the northern polar cap. Figure 33 compares average absolute values of the vertical component $|Z|$ at Vostok ($-89°$) near the southern magnetic pole with those at Resolute Bay ($83°$). These averages are calculated in each IMF sector using observations during 8 to 10 h in the geomagnetic local time, and three periods in 1964 (corresponding to three seasons) are examined. The open and filled boxes on the abscissa show the prevailing direction of IMF: the open boxes refer to

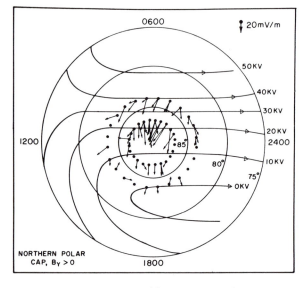

(a)

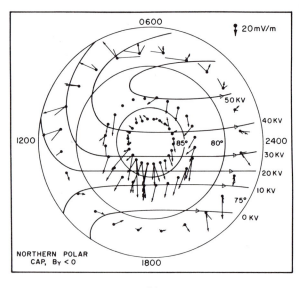

(b)

Fig. 32a,b. Hourly average electric field vectors for the data collected (a) when $B_y > 0$ and (b) when $B_y < 0$ (Mozer et al., 1974). Arrows on equipotential contours indicate the direction of convection

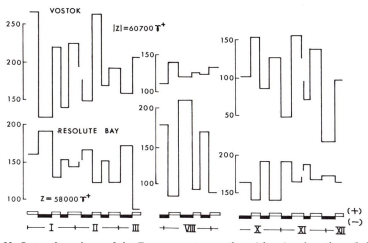

Fig. 33. Sector dependence of the Z-component at southern (above) and northern (below) polar-cap stations. The sector polarity is indicated by boxes on the abscissa (Mansurov, 1969)

the away sector (associated with eastward B_y) and the filled boxes refer to the toward sector (associated with westward B_y). The polarity dependence observed at Resolute Bay (where $|Z| = Z$) is exactly that noted in Figure 28a. In contrast, the dependence observed in $|Z|$ at Vostok is opposite to it. Since an increase in $|Z|$ means presence of an upward perturbation field at Vostok, the observed dependence can be translated to mean that the zonal current around the southern magnetic pole is directed westward in the away sector and is eastward in the toward sector: namely, opposite to what is observed in the northern polar cap in the same IMF sector (Mansurov, 1969). The corresponding hemispherical asymmetry has been noted in the satellite observations of the electric field; when B_y is eastward (westward) and the dawn-to-dusk electric field in the northern hemisphere is stronger on the dawn (dusk) side of the pole, the same electric field is stronger on the dusk (dawn) side of the southern pole (Heppner, 1972).

The hemispheric asymmetry that is dependent on the solar-magnetospheric east–west component B_y of IMF has been detected also in the auroral-zone magnetic activity; the activity tends to be higher in the northern (southern) auroral zone when B_y is directed eastward (westward) (Yoshizawa and Murayama, personal communication). The effect is predominant around midnight, while in the polar cap the B_y dependence has been observed to stand out around midday (Berthelier and Guerin, 1973). Correspondingly, the geomagnetic activity in the northern auroral zone is higher when B_y is eastward than when it is westward (Murayama and Hakamada, 1975). The nature of this effect has yet to be clarified.

Annual and UT Variations in Geomagnetic Activity

In addition, the auroral-zone magnetic activity shows the IMF-polarity effect that appears equally in both hemispheres. This effect depends on the time of a year and constitutes a substantial part of the semiannual variation of the geomagnetic activity. Figure 34a shows the annual variation of the Am index, which is obtained by taking the average of An and

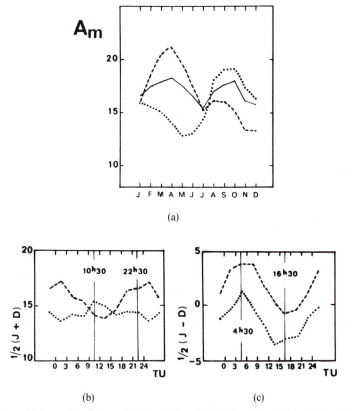

(a)

(b)   (c)

Fig. 34a–c. (a) Annual variation of the Am index, (b) UT diurnal variation of Am, the June and December data being averaged, and (c) UT diurnal variation of Am, the December data being subtracted from the June data. (———) Averages for all days; (· · · ·), (----) averages made separately over days of away and toward polarity, respectively (Berthelier, 1976)

As indices representing the activities in northern and southern hemispheres[7]. The data are divided into three groups: namely, all data (solid lines), data obtained during intervals of the IMF away polarity (dotted lines) and of the IMF toward polarity (dashed lines). In Figure 34a we can see the following: (1) while, on the average, Am shows a semiannual variation with maxima in equinoctial months, (2) the spring maximum is predominant during the toward polarity and the fall maximum is predominant during the away polarity. Point (2) may appear to indicate that there is a B_y effect having a hemispheric symmetry, but this feature has been found to be attributable to the dependence of the auroral-zone magnetic activity on the IMF southward component noted in the previous Section. The effect of the southward component is involved in this apparently polarity-dependent effect because there is difference between the solar rotational axis and the geomagnetic axis. In the framework where z axis is the solar rotational axis, average IMF has a positive eastward (westward) component in the away (toward) sector but zero B_z in each sector. When viewed from the framework fixed to the geomagnetic dipole (such as the solar magnetospheric coordinates), however, such an average field is generally seen to have a finite north–south component. The sign of this north–south component is dependent both on the polarity of IMF and on the angle between the solar axis and the geomagnetic axis, and it can be shown that the largest contribution to the southward component is made in October (April) in the away (toward) sector. Since the magnetic activity in the auroral zone is greater when the southward IMF (relative to the geomagnetic axis) is larger (see Fig. 27), the polarity-dependent annual variation [i.e., point (2)] is expected to result. Note that the B_y component of IMF in the solar magnetospheric coordinates is not involved in the process. The idea has been substantiated by quantitative comparison with the observation (Russell and McPherron, 1973a).

The preceding model predicts further that the geomagnetic activity would be dependent on UT, because on any day of the year the angle between the solar axis and the geomagnetic axis varies with UT. The UT diurnal variation of this angle is such that in the away (toward) sector the spiral field in the solar equatorial plane makes a maximum contribution to the solar-magnetospheric southward component (and

---

[7] In order to scale the level of the geomagnetic activity in each hemisphere, Mayaud (1967) produced the 3-hourly An and As indices, which represent the activity in the northern and the southern hemisphere, respectively. These indices are based on the K index that represents the range of the magnetic perturbation observed at individual stations, and K indices from 10 (6) stations, most of which are located in the geomagnetic latitude range of $45°$ to $55°$ ($-45°$ to $-55°$), are used for constructing An (As). The index Am, the measure of the global activity, is derived by averaging An and As.

hence increases the geomagnetic activity) around 1030 (2230). This prediction also seems to be consistent with the observation reproduced in Figure 34b, where the average of Am diurnal variations observed in June and December solstices is plotted against UT (Berthelier, 1976).

Then it might appear that the point (1) noted in Figure 34a is attributable to the superposition of the annual variations corresponding to different IMF polarities. However, the amplitude and the detailed phase of the observed semiannual variation indicate that there should be other sources. Hence Murayama (1974) suggested that the heliographic latitude dependence of the solar wind velocity and the IMF strength, both of which appear to increase with increasing latitude on the average, should be taken into account; increases in these parameters augment the strength of the interplanetary electric field. On the other hand, Berthelier (1976) supported the earlier hypothesis that the geomagnetic activity is enhanced when the angle between the geomagnetic axis and the earth–sun line becomes close to 90°. As evidence in favor of this conjecture, she produced Figure 34c, which shows the polarity-independent part of the UT variation of the geomagnetic activity. (The apparent polarity effect noted earlier has been eliminated by taking the difference between the two solstitial observations; according to the Russell-McPherron model the polarity effect is expected to produce equal perturbation fields in summer and winter. Indeed the UT variation of Figure 34c is essentially independent of the sector polarity.) Note that 0430 UT in June and 1630 UT in December, when the activity is found to maximize, are the times when the preceding angle comes closest to 90°. The theoretical basis of this hypothesis has been sought in the suggestion that the solar wind energy is imparted to the magnetosphere by the Kelvin-Helmholtz instability at the magnetopause; as will be shown in Section V.3, the threshold value of the solar wind velocity required to excite this instability is minimum when the preceding angle is 90° (Boller and Stolov, 1970).

II.4 The Open Model of the Magnetosphere

Since the penetration of the interplanetary electric field E_I via connected field lines into the magnetosphere is necessarily associated with the generation of the bulk motion of the magnetospheric plasma, the open magnetosphere is intrinsically a dynamic subject. Geomagnetic field lines are also convected with plasma in which they are frozen, and they change their configuration from the 'closed' type (having both feet on the earth) to the 'open' type (having only one foot on the earth) as they are cut and 're' connected when they make contact with interplanetary field lines on the day-side magnetopause. The point where the reconnection occurs is considered to form a line, and this line is called 'reconnection line.'

The Reconnection Line

If the direction of the interplanetary magnetic field were perfectly aligned to the dipole moment of the earth, the reconnection line would be a neutral ring that surrounds the earth in the geomagnetic equatorial plane, and the day-side reconnection would occur between field lines having precisely opposite orientations (Fig. 18). In reality, however, IMF directions are seldom aligned to the geomagnetic dipole. Although the superposition of arbitrary IMF on the geomagnetic field would still produce neutral points scattered here and there on the magnetopause, the reconnection at such neutral points would not have a practical influence on the gross dynamics of the magnetosphere. This is because the amount of the magnetic flux that gets reconnected on the day-side magnetopause per unit time is proportional to the length of the reconnection line, and it becomes zero if the reconnection line becomes a point which, by definition, has zero length. In other words, the potential difference of the E_I field introduced into the magnetosphere is proportional to the lateral width of the reconnecting IMF field lines, and it vanishes when the reconnection line is reduced to a point.

Hence in actual circumstances where the IMF direction is seldom due southward, the day-side reconnection should occur on a reconnection line, which is not characterized by the neutrality of the magnetic field, if the reconnection mechanism is to be maintained as the principal mechanism of energy transfer from the solar wind to the magnetosphere. The problem then is to reconsider the physics of the reconnection process and find out where on the day-side magnetopause the reconnection line is formed under general IMF orientations. There seem to be at least two conditions that the reconnection line should satisfy. The first is that the reconnection line should pass the stagnation point of the solar-wind flow in contact with the magnetopause. This condition is derived from the consideration that if the reconnection line were not 'hooked' to the stagnation point it would be carried away with the solar wind and could not operate steadily on the day-side magnetopause.

[It is noted that the occurrence of the reconnection would not cause significant modification (such as the reversal of the velocity) to the flow pattern of the solar wind around the magnetosphere. This is because the deflection of the flow by the $J \times B_N$ force (where J is the surface current flowing on the magnetopause and B_N is the component of the magnetic field normal to the magnetopause) is insignificant if $|B_N| \ll |B_T|$ (where B_T is the tangential component of the magnetic field immediately inside the magnetopause), since the momentum flux of the solar wind is comparable to $J \times B_T$. According to observations $|B_N|$ seems to be indeed less than $|B_T|$ on the day-side magnetopause (Sonnerup and Ledley, 1974).]

The day-side nose of the magnetosphere is obviously a stagnation point of the solar wind flow, and the first condition is satisfied for a line that passes through this point. The second condition on the reconnection line has to be the one that specifies its direction. In an idealized circumstance where the interplanetary and the geomagnetic fields in contact at the magnetopause have the same strength and differ only in their directions, symmetry considerations suggest that a direction halfway between the two field directions is the direction of the reconnection line. The reconnection line derived from the foregoing conditions is illustrated schematically in Figure 35. In this Figure, the front side of the doughnut represents the outer surface of the closed field lines in the dayside magnetosphere and the dashed line represents the reconnection line. Figures 35a and b correspond to the cases of IMF having southward and northward component, respectively. The width L refers to the lateral dimension of the IMF fluxes that are reconnected, and the electric potential difference that is introduced into the magnetosphere is the product of the interplanetary electric field E_I and L (Nishida and Maezawa, 1971).

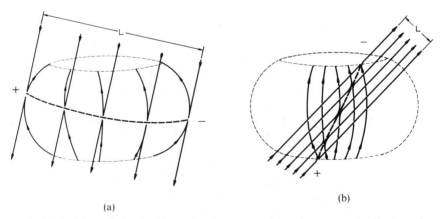

(a)   (b)

Fig. 35a,b. Schematic drawing illustrating the reconnection at the reconnection line (----) passing the nose of the magnetopause. IMF is supposed to be oriented south–west in (a) and north–east in (b). The polarity of the electric field is indicated by + and − signs (Nishida and Maezawa, 1971)

To extend the preceding model to more general circumstances where magnetic fields on the interplanetary and the geomagnetic sides of the magnetopause do not have the same strength, we note the following: the direction halfway between the two fields having the same strength is the direction of the electric current J flowing on the interface. Since the Lorentz force acting on such J has no component parallel to the line, two branches of open field lines created by the reconnection move apart, without being dragged parallel to the reconnection line. Gonzalez and Mozer (1974) and also Sonnerup (1974) adopted this feature, namely the parallelism to the direction of the magnetopause current as the second condition characterizing the reconnection line.

Denoting the magnetic fields on interplanetary and geomagnetic sides of the magnetopause by B_1 and B_2 and adopting a coordinate system where the flow velocity v is directed toward the y–z plane from both sides, they illustrated the situation (Fig. 36). The y axis, which is perpendicular to $\Delta B \equiv B_2 - B_1$, gives the direction of the reconnection line ($B_2 \geq B_1$ is assumed.) The components of B_1 and B_2 that are normal to the reconnection line are

$$B_1 \sin (\alpha - \beta) = \frac{B_1(B_1 - B_2 \cos \alpha)}{(B_1^2 + B_2^2 - 2B_1B_2 \cos \alpha)^{1/2}} \equiv B_1 F_1 \left(\frac{B_1}{B_2}, \alpha \right) \quad (51)$$

and

$$B_2 \sin \beta = \frac{B_2(B_2 - B_1 \cos \alpha)}{(B_1^2 + B_2^2 - 2B_1B_2 \cos \alpha)^{1/2}} \equiv B_2 F_2 \left(\frac{B_1}{B_2}, \alpha \right) \quad (52)$$

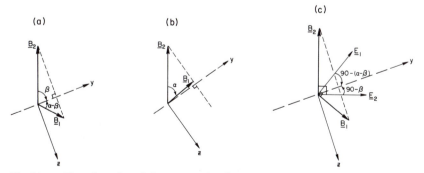

Fig. 36a–c. The orientation of the reconnection line (y axis) with respect to magnetic fields (B_1 and B_2) and electric fields (E_1 and E_2). Subscripts 1 and 2 refer to different sides of the interface. The reconnection occurs in (a) but not in (b), and (c) illustrates the electric fields for case (a) (Gonzalez and Mozer, 1974)

As shown in the Figure, α is the angle between the two fields $\boldsymbol{B}_1$ and $\boldsymbol{B}_2$, and β is the angle between $\boldsymbol{B}_2$ and the reconnection line. For the reconnection to occur, B_1F_1 and B_2F_2 should have the same sign. Since B_2F_2 is always positive as $B_2 \geq B_1$, the necessary condition for the reconnection is hence $B_1 > B_2 \cos \alpha$. Figure 36a illustrates a case where this condition is satisfied. When $B_1 < B_2 \cos \alpha$, on the other hand, the field components perpendicular to the reconnection line are parallel to each other and there can be no reconnection (Fig. 36b). Note that the reconnection can always occur when the solar magnetospheric B_z component of IMF has a southward polarity, since $\alpha > \pi/2$ and $\cos \alpha < 0$, but when $\boldsymbol{B}_z$ has a northward polarity the occurrence of reconnection is limited to a range defined by $\alpha > \cos^{-1} (B_1/B_2)$. In the following discussion $B_1 > B_2 \cos \alpha$ will be assumed except when noted otherwise.

The electric fields given by $\boldsymbol{E} = -\boldsymbol{v} \times \boldsymbol{B}$ are also shown in Figure 36c. If both the B_1/B_2 ratio and the α angle do not vary much on the entire day-side magnetopause, the electric potential drop along the reconnection line can be estimated by

$$\Phi = (E_1 \sin (\alpha - \beta))W \tag{53}$$

$$= v_1 B_1 W F_1 \left(\frac{B_1}{B_2}, \alpha \right)$$

or equivalently

$$\Phi = (E_2 \sin \beta)W \tag{54}$$

$$= v_2 B_2 W F_2 \left(\frac{B_1}{B_2}, \alpha \right)$$

where W is the length of the reconnection line and WF_1 corresponds to L of Figure 35. As seen in Figure 35 the electric field along the day-side reconnection line of this type is directed from the dawn side to the dusk side regardless of whether B_z of IMF is southward or northward, and so is the electric field produced over the polar cap by projecting the preceding electric field along open field lines. The equality of (53) and (54) is to be understood as a condition relating the velocities v_1 and v_2.

In reality, however, both $\boldsymbol{B}_1$ and $\boldsymbol{B}_2$ vary their magnitude and direction over the day-side magnetopause so that neither the B_1/B_2 ratio nor the α angle takes a fixed value over the entire day-side magnetopause. Consequently the angle β characterizing the local direction of the reconnection line relative to the local geomagnetic field $\boldsymbol{B}_2$ is not a constant quantity. Hence it is necessary to assess the required corrections for the expressions (53) and (54) by modeling the spatial variations of $\boldsymbol{B}_1$ and

B_2. Gonzalez and Mozer (1974) used a simple hemispheric representation for the day-side magnetopause and derived the following expression:

$$\Phi = \frac{2v_sB_{I,t}RF_1\left(\dfrac{KB_{I,t}}{B_G}, \alpha_0\right)}{\sin(\alpha_0/2)} \tag{55}$$

where v_s is the speed of the solar wind, $B_{I,t}$ is the tangential component $[(B_y^2 + B_z^2)^{1/2}]$ of the interplanetary magnetic field, R is the radius of the day-side magnetopause, $KB_{I,t}/B_G$ is the ratio of interplanetary and geomagnetic fields at the nose of the magnetopause (K being the amplification factor of the field at the bow shock), and α_0 is the value of the α angle at the nose of the magnetopause. The $\sin^{-1}(\alpha_0/2)$ factor in Eq. (55) reflects the fact that the day-side magnetopause is more oblate than a sphere, namely, its radial distance increases as one departs from the nose at the subsolar point. (This oblateness was expressed in their model by placing the center of the hemisphere $5R_E$ tailward of the dipole position.)

Figure 37 is a comparison of the electric field calculated from the preceding formula with that observed in the polar cap. The measured field is the dawn-to-dusk component of the electric field observed by balloons at Resolute Bay and Thule. The calculation of Φ is made by using the following parameters in addition to the simultaneous IMF observations: $B_G = 70\gamma$, $R = 15R_E$, $K = 5$, $v_s = 300$ km/s, and Φ is reduced to the 'model field' by assuming the size of the polar cap to be 3000 km wide. The model field is equated to zero when $KB_{I,t} \leq B_G \cos \alpha_0$. A close similarity found between the time variations of the measured and

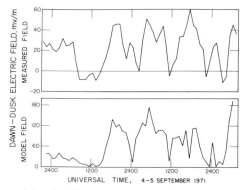

Fig. 37. Comparison of electric fields measured by balloons in the polar cap (top) and calculated by a model (bottom) (Gonzalez and Mozer, 1974)

the modeled fields gives support to the foregoing estimate of Φ, but there is a factor difference of $\sim\frac{1}{3}$ in their magnitudes. This $\frac{1}{3}$ factor can be interpreted to represent the fraction of the interplanetary field lines that can reach the reconnection line; among the IMF flux tubes of width L that approach the reconnection line, only $L/3$ can make it, the remaining $2L/3$ being deflected and pushed aside to the flanks of the magnetosphere without ever touching the reconnection line. That the observed fraction is less than one means that there is a limit in the reconnection rate that can be achieved. [According to Petschek's (1964) reconnection model the upper limit is about 10% of the Alfven speed in the undisturbed upstream flow.]

While the electric potential given by Eq. (55) involves both B_y and B_z components of IMF, the majority of the correlational studies between the IMF and the geomagnetic field has been carried out separately for either the B_y or B_z component. If Φ given above is the proper representation of the electric field resulting from the reconnection process, it has to be explained why apparently good correlations have been found between B_z and DP2, and also between B_z and substorm activities. For this purpose, let us rewrite Eq. (55) in terms of components of $\boldsymbol{B}_I$;

$$\Phi = \frac{2^{3/2} v_s B_{I,t}^{1/2} R(KB_{I,t}^2 - B_G B_{I,z})}{(B_{I,t} - B_{I,z})^{1/2}(K^2 B_{I,t}^2 + B_G^2 - 2KB_G B_{I,z})^{1/2}} \tag{56}$$

where $B_{I,t} = (B_{I,y}^2 + B_{I,z}^2)^{1/2}$ and $B_{I,t}\cos\alpha_0 = B_{I,z}$. This equation applies to the case of $KB_{I,t}^2 > B_G B_{I,z}$, and Φ is zero when $KB_{I,t}^2 \leq B_G B_{I,z}$. Φ is plotted in Figure 38 versus B_z ($\equiv B_{I,z}$) using $|B_y|$ ($\equiv |B_{I,y}|$) as a parameter. It can be seen that Φ increases with an increase in $(-B_z)$ at any fixed value of $|B_y|$, and that the variation in Φ is much more sensitive to $|B_z|$

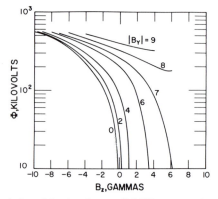

Fig. 38. A model calculation of the electric potential difference produced along the reconnection line. B_y and B_z are components of IMF (Gonzalez and Mozer, 1974)

than to $|B_y|$ when B_z is in the southward range. Hence it seems possible that the reported correlations with southward B_z are in fact reflections of the dependence on Φ.

However, the problem of the modeling of the day-side reconnection should not be considered to have been settled. The attitude of the reconnection line that has been assumed in the foregoing models has not yet been verified by in situ measurements of the field topology; in fact the existence of the reconnection line itself still awaits observational confirmation. Further improvements of the available model would be warranted as the correlational analyses between IMF and geomagnetic data yield more advanced results.

The preceding estimate of Φ is concerned solely with the total potential drop introduced into the magnetosphere, and the spatial distribution of the potential inside the polar cap is not yet specified. When $B_y = 0$ the magnetic-field structure as well as the electric-field distribution of the open magnetosphere should be symmetric with respect to the noon–midnight meridian. In that circumstance the electric equipotential contours at ionospheric heights are expected to form two symmetrical vortices (in each hemisphere) that are centered on dawn and dusk edges of the polar cap. These equipotentials represent the streamlines of the convective motion, and the flow direction is symmetric with respect to the noon meridian, being eastward before midday and westward after midday outside the polar cap. When B_y differs from zero and the reconnection line is inclined with respect to the equatorial plane, however, the Lorentz force acting on the open field lines has a longitudinal component, and this would produce zonal flow of field lines on the day-side edge of the polar cap. This can be seen in Figure 39; Figure 39a is the schematic illustration of the front side of the magnetosphere when $KB_{I,t}$ has only the eastward component and is equal in magnitude to B_G. The reconnection line is labeled L'L, and thick arrows indicate the direction of the Lorentz force $\boldsymbol{J} \times \boldsymbol{B}_N$ and the resulting separation of the open field lines. Figure 39b is the view of field lines from the north, and thick arrows indicate their motion. The reconnection with the eastward IMF causes field lines at the day-side edge of the northern polar cap to be convected zonally toward the west. Consequently, the size of the convection vortex (and equally the equipotential vortex) is expected to be enlarged on the dusk side but reduced on the dawn side. For the same reason the eastward zonal flow would be generated and the dawn-side vortex would become larger than the dusk-side one when B_y has the westward component. In the southern hemisphere the effect is expected to work in the opposite way.

When the preceding expectation is compared with observed directions of the B_y-dependent circumpolar currents and observed asymmetries of

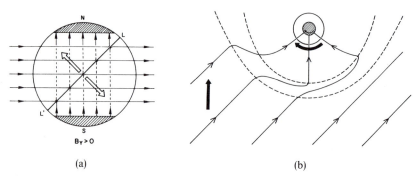

Fig. 39a,b. Dayside reconnection with IMF having eastward polarity. (a) is the view of the reconnected field lines from the sun, and (b) is the view from the north (Mozer et al., 1974). In both parts, thick arrows indicate the direction of motion of the newly reconnected field lines

vortex sizes, an essential agreement can be seen. (Note, however, that convection streamlines of Figure 32 are meant only to represent the gross structure, and hence the detailed flow pattern at the polar-cap boundary is not necessarily precisely expressed.) A characteristic feature of the polar-cap electric field predicted by the 'reconnection line' model is that there is no net difference between average values of electric potentials of the northern and the southern polar caps. All the electric equipotentials entering both polar caps originate from the common reconnection line, and the north–south asymmetry in the electric field arises only from the way they spread in the respective polar caps. A net difference between northern and southern polar-cap potentials would have been expected if it were assumed that the ionosphere was exposed directly to the E_z component of the interplanetary electric field given by $-\boldsymbol{v}_s \times \boldsymbol{B}_y$.

Magnetopause and Polar Cap in the Open Magnetosphere

In the open magnetosphere transition of field and particle characteristics from the magnetospheric ones to the solar-wind ones does not occur at a single discontinuity surface but is accomplished at a combination of wave fronts. This is because the information that the different plasma regimes have made contact is transmitted into both regimes as hydromagnetic waves of Alfven and slow modes that travel along magnetic lines of force; the connected field lines act as guides that spread the information. The information (namely these waves) starts to travel when the field lines become reconnected at the reconnection line, and on the magnetospheric side it propagates toward the polar cap. A detailed examination of the observation in the light of theoretical models that incorporate the preceding consideration [first suggested by Petschek

(1966)] is yet to be conducted, but plasma observations in the high-latitude magnetosphere seem to compare nicely with the essential prediction of the open magnetosphere. The observations concerned are those of protons that are detected in the region designated as plasma mantle in Figure 40. The velocity of these protons have a component directed away from the sun and they seem to represent the solar-wind plasma that has entered the magnetotail along the connected field lines (Rosenbauer et al., 1975).

Full and open circles in Figure 40 show schematically the expected trajectories of relatively more energetic and less energetic of these protons; the slower particles are displaced further toward the antisunward direction as they are influenced more by the antisunward magnetospheric convection during its longer stay inside the magnetosphere. Observations of the locations of low-latitude limits of H^+, He^{2+}, and electron precipitations at 800-km altitude indeed show the expected dependence on the particle speed (Shelley et al., 1976). The intimate relation between the mantle formation and the openness of the magnetosphere is suggested further by the high correlation that has been found between the detection of the plasma mantle and the southward polarity of IMF (Sckopke et al., 1976).

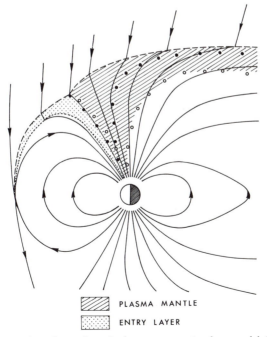

Fig. 40. Plasma mantle and entry layer in the open magnetosphere model. (————) the slow expansion fan

Plasma with the magnetosheath characteristics has been observed not only on the tail field lines but also on the day-side field lines located equatorward of the field line passing the cusp. The latter plasma has been designated by Paschmann et al. (1976) as entry layer plasma. According to our concept of the magnetopause structure in the open magnetosphere, the entry layer is expected to be bounded on the low-latitude side by the slow wave extending from the reconnection line. In order to give a quantitative explanation to plasma observations in the entry layer and the plasma mantle, account would have to be taken of such factors as the particle mirroring and wave reflection at ionospheric heights, the diffusion across field lines under fluctuating electromagnetic fields, and the variability of the IMF polarity. The occurrence of the diffusion process is suggested by the coexistence [reported by McDiarmid et al. (1976)] of the trapped electrons (with $\sim 90°$ pitch angle at altitudes of ~ 1400 km) with the plasma having the magnetosheath characteristics, and is supported by the observed energy dispersion (namely, the variation of particle energy with latitude) of such plasma (Reiff et al., 1977).

Another important property of the open magnetosphere model is that the radius of the day-side magnetosphere and the size of the polar cap vary with IMF. This is because the interplanetary electric field drives the current I_L, which flows along open field lines and closes via ionospheric Pedersen current. The geometry is illustrated in Figure 41. The magnetic field ΔB produced by this current system acts to reduce the geomagnetic field in the vicinity of the day-side magnetopause. (The tail current I_T also has the same effect but it is quantitatively less important.) The intensity of the I_L current system is governed by the ionospheric Pedersen conductivity and the electric field produced in the polar cap. The circuit may be regarded as a shunt introduced into the interplanetary electric field due to the contact with the ionosphere through the open field lines. Since the interplanetary electric field as seen by the magnetospheric observer is the consequence of the solar wind motion relative to the interplanetary magnetic field, the dissipation of the electric-field energy by the shunting circuit is to be compensated by the deceleration of the solar-wind flow. In the region where the deceleration occurs there exists so-called inertia current j_A given by

$$j_A = \frac{nm}{B^2} \left[\boldsymbol{B} \times (\boldsymbol{v} \cdot \nabla)\boldsymbol{v} \right]. \tag{57}$$

The inertia current, which closes the current system on the solar wind side, would flow in the vicinity of the magnetopause so that it is depicted as I_M in Figure 41.

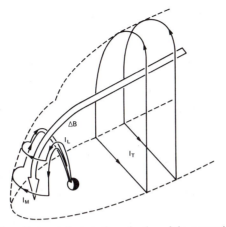

Fig. 41. Current systems that contribute to the reduction of the magnetic flux in the dayside magnetosphere. I_L and I_T: line-tying and cross-tail currents; and I_M: the closure of the line-tying current in the neighborhood of the magnetopause (Maezawa, 1974)

The I_L current system is sometimes referred to as the line-tying current because it arises from the property of the ionosphere to dissipate the electric field energy (by the Pedersen conductivity resulting from the collision of charged particles with neutral constituents) and retard the convective motion of plasma and field lines. The presence of the line-tying current has been confirmed by satellite observations of the magnetic field at the height of 800 km; the current flows into the ionosphere in the morning sector and away from the ionosphere in the evening sector, the maximum current densities being observed during $7 \sim 8$ LT and $15 \sim 16$ LT, respectively, and the total current intensity during slightly disturbed conditions is about 10^6 A (namely, about 3×10^{15} statamp) (Iijima and Potemra, 1976a). Since the scale of the I_L–I_M current loop is approximately $10R_E$, the magnetic flux produced by the circuit is estimated to be of the order of 10^8 Wb. This results in an appreciable reduction of the magnetic flux in the day-side magnetosphere, and the magnetopause has to be brought closer to the earth in order to achieve the same magnetic pressure as is obtained in the absence of the line-tying current system. Hence the magnetosphere is expected to shrink and the day-side boundary of the polar cap is expected to move equatorward as the potential difference introduced into the magnetosphere by reconnection is increased, even when the dynamic pressure of the solar wind is kept the same.

The IMF effect on the position of the day-side magnetopause was first witnessed in one fortuitous occasion when a satellite (OGO 5) during an inbound pass closely traced the receding magnetopause for an interval of more than two hours. Since the dynamic pressure of the solar wind did not vary, the 'erosion' of the magnetopause was interpreted to be due to the southward orientation of B_z that started at the beginning of the interval and lasted throughout (Aubry et al., 1970). The statistical confirmation that the southward B_z brings the magnetopause closer in as compared to when B_z is northward has since been made, and the loss of the day-side flux is found to be about 10^8 Wb (Maezawa, 1974). The associated enlargement of the polar cap has been detected by the variation of the latitudinal position of the low-energy (~ 1 keV) electron precipitation in the day sector (9–15 h). Filled circles of Figure 42 represent the latitude of the equatorward boundary of this precipitation and are plotted against the average of B_z during the preceding 45-min interval. It is seen that the latitude is around 76° when B_z is northward, but it becomes several degrees lower than this when B_z is southward. The rate of the equatorward shift of the boundary following a sharp southward turning of B_z is about 0.1°/min (Burch, 1973). The open circles of Figure 42 representing the poleward boundary of the precipitation show nearly the same trend as the filled circles. The B_z dependence exists also in the size of the polar cap current system as noted in Figure 29; this is understandable because distributions of electric fields and currents generated by the reconnection process are fixed relative to the boundary of the polar cap.

Finally we have to discuss the response time of the magnetosphere to variations in the IMF condition. The magnetospheric reconnection with the variable IMF seems to have at least four characteristic time scales.

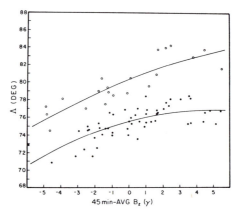

Fig. 42. Plots of poleward (○) and equatorward (●) boundary latitudes of the ~ 1 keV electron precipitations against the 45-min average of the IMF north–south component (Burch, 1973)

The first, τ_1, is the time in which the change of the state of the day-side reconnection line is communicated to the ground. It would be given roughly by s/V_A, where s is the length of the geomagnetic field line reaching the reconnection line and V_A is the Alfven speed. τ_1 would thus be comparable to the transmission time of the SI from the magnetopause to the ground, and it is estimated to be about 1 min. The second time scale, τ_2, arises from the motion of the magnetopause in response to IMF variations. When the magnetopause is moving with a velocity v_b, the interplanetary electric field as seen by the reconnection line located on it is not $E_1 = -v_s \times B_1$ but $-(v_s - v_b) \times B_1$; the electric field applied on the reconnection line differs from E_1 until the line-tying current reaches the equilibrium intensity and the motion of the day-side magnetopause stops. Coroniti and Kennel (1973) estimated τ_2 on the assumption that the total line-tying current intensity is equal to the decrement of the total intensity of the magnetopause current (flowing on the day-side magneto-pause between polar cusps) due to the reduction in the area of the day-side magnetopause, which, in turn, is brought about by the presence of the line-tying current. According to them $\tau_2 \sim 400 \, \Sigma_P$, where Σ_P is the Pedersen conductivity (in mhos), and it has been estimated to be around ten minutes.

The third time scale, τ_3, is considered to reflect the time that is required for the attitude of the reconnection line to adjust to the new state of IMF. It includes also the correction $-v_r \times B_1$ to the electric field that arises from the motion (with velocity v_r) of the reconnection line on the magnetopause. This time scale does not seem to have been assessed, however. The fourth time scale, τ_4, is the length of time during which the IMF field lines, once reconnected, are kept connected to geomagnetic field lines until the connection is terminated by another reconnection process operating in the magnetotail. If E_1 is communicated equally along all these connected field lines, the effective electric field that appears in the magnetosphere would be the average of the fields that have been imposed on the reconnection line over an interval τ_4 before present (Nishida and Maezawa, 1971; Holzer and Reid, 1975). $\tau_4 \sim D/v$, where D is the diameter of the polar cap and v is the noon–midnight drift speed across the polar cap. Since the drift speed corresponding to the average polar cap electric field of 40 mV/m (1×10^{-6} cgs) is 0.6 km/s, τ_4 is estimated to be 5×10^3 s using $D \sim 3000$ km.

When compared with the delay time of ~ 15 min found between IMF and its geomagnetic effects, τ_1 is much smaller and τ_2 is comparable. τ_4, however, is appreciably longer, indicating that the effective value of D is only ~ 500 km. Hence it appears that E_1 is communicated only along those open field lines that are rooted in the vicinity of the day-side edge of the polar cap and not along all the open field lines defining the polar cap. This is probably related to the fact that the night-side reconnection process occurs independently of the day-side one so that the open field lines convected from the day side are piled up in the tail for some time; when the field configuration is changing, the induction electric field is produced, and magnetic field lines connecting the polar cap to the distant solar wind via magnetotail do not act as equipotentials of the electric field.

II.5 Reconnection to Northward IMF, and S_q^p

Correlation with Northward IMF

The reconnection line passing the day-side nose of the magnetosphere functions under any IMF that has a southward component, but it ceases to operate when IMF has a northward component that exceeds $KB_{I,t}^2/B_G$.

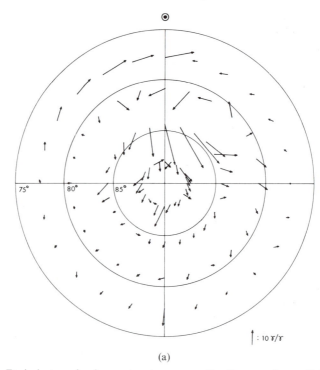

(a)

Fig. 43a. Equivalent overhead current vectors representing the regression coefficient of the polar-cap magnetic field on B_z for the case of $B_z > 1\gamma$. The length of the unit vector shown at bottom right corresponds to a 10γ increase in the horizontal geomagnetic field produced by a 1γ increase in the IMF northward component

(For typical conditions of IMF, $KB_{I,t}^2/B_G \sim 1$ to 2γ.) If the reconnection line passing the day-side nose represents the only mode of the magnetospheric interaction with IMF, it would hence be expected that when IMF has a large northward component only the quiet daily variation Sq remains in the polar cap. It was noted in Figure 29, however, that when B_z is positive there appears in the summer polar cap another mode of the disturbance field that is different from both Sq and DP2.

The nature of this mode is pursued in Figure 43a where the partial regression coefficients on B_z obtained in the $B_z > +1\gamma$ range are plotted in the form of the equivalent overhead current vectors. The material used and the method employed are the same as those with which Figure 24 was constructed, the only difference being in the range of B_z. The influence of the B_y component has been eliminated, and the separation from the background Sq variation has been achieved through the use of the multiple regression method. Figure 43b is the equivalent overhead current system that is derived from the current vectors of Figure 43a. The obvious difference between this current system and Figure 24, which corresponds

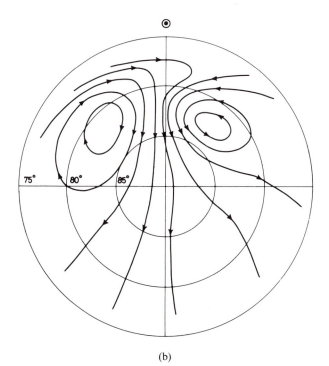

(b)

Fig. 43b. Equivalent current pattern in the polar cap derived from (a). The total current induced between two consecutive streamlines by a 1γ increase in B_z is about 4×10^3 A (Maezawa, 1976)

to the southward B_z (or to $B_z < +1\gamma$, to be more exact), is that the direction of the current flow over the pole is reversed; the current is directed from day to night when $B_z > +1\gamma$, while it is from night to day when $B_z < +1\gamma$. Furthermore, the foci of the equivalent current vortices are located at significantly higher latitudes and shifted toward the day side in the present current system. Hence the equivalent current system produced by the northward B_z is not simply the reversal of the current system produced by the southward B_z. This observation has led to the conclusion that the interaction between the IMF and the magnetosphere has one more mode that is distinct from the nose reconnection process discussed in the last Section (Maezawa, 1976).

The reversal of the equivalent current flow over the polar cap has been detected in some individual cases when B_z is northward. Figure 44 is a comparison of records of the B_z component of IMF, D and H component records at Vostok ($-89°$), and the AE index. The dot-dash curve labeled as DS-phase indicates the sign of the perturbation field at Vostok that is produced by the transpolar current directed roughly from night to day. There are several intervals (shaded) where the sign of the observed perturbation at Vostok is different from what is expected from such a current, and they are seen to correspond to intervals of northward B_z (Iwasaki, 1971).

In the polar cap where the electric conductivity would not have pronounced small-scale nonuniformities, the streamlines of the equivalent current system such as Figure 43b can be considered to bear close resemblance to the equipotential contours of the driving electric field. This view

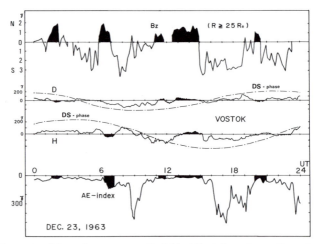

Fig. 44. Response of the polar-cap magnetic field at Vostok on the northward component B_z of IMF (Iwasaki, 1971)

is based on model calculations of the $j - E$ relation such as Lyatsky et al. (1974) (Fig. 25). Hence the magnetospheric convection generated by the interaction with the northward IMF can be inferred approximately by reversing the direction of the equivalent current flow of Figure 43b.

It can be seen that the convection of this mode has two major characteristics: first, the field lines are convected from the night side to the day side over the magnetic pole, and second, the convection is confined to the high latitude region above $\sim 75°$. The second point indicates that the convection is confined to the polar cap and does not involve closed field lines. The reconnection process that produces this type of convection is illustrated in Figure 45. When an IMF field line having northward B_z (labeled 2′–2) makes contact with the magnetopause at the surface of the magnetotail, it becomes connected with a tail field line labeled b. Of the two branches of field lines thus produced, the branch c–3 has its root on the earth and thus belongs to the class of open field lines of the magnetosphere. It is convected to d–4, e–5, and eventually returns to b after passing the central portion of the tail like a. As the field line in the northern polar cap is connected to the interplanetary field line coming from the south, its convection toward the night side, which may be visualized to result from the dragging of its interplanetary end by the solar-wind flow, is forced to occur at the edge of the polar cap. Hence the convection is directed from day to night at the edge of the polar cap and the return flow directed from night to day appears over the pole. Another branch of the field lines produced by the reconnection forms a purely interplanetary

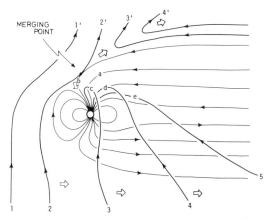

Fig. 45. Field and flow pattern around the magnetosphere when northward IMF lines are reconnected with tail field lines. The Figure represents the process proceeding on the evening side of the northern polar cap only, and a similar process is thought to occur on the morning side and in the southern polar cap (Maezawa, 1976)

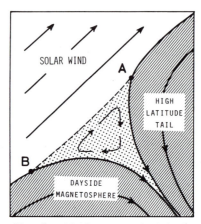

Fig. 46. Trapped plasma (dotted) and its flow pattern in the polar cusp. *A* and *B* represent stagnation points of the trapped plasma flow (Maezawa, 1976)

field line that is not rooted to the earth and is swept away with the solar wind as field lines labeled 3′ and 4′. Although omitted from this Figure, a similar process is thought to occur in the southern lobe of the tail as well. The convection of this mode, which involves only open field lines in the magnetotail, was envisaged by Russell (1972) prior to the detailed observation of the northward IMF effect.

The foregoing model implies that reconnection lines having finite longitudinal extents are formed on both northern and southern surfaces of the magnetotail. This requires that there are on the tail surface stagnation points of the solar-wind flow; unless the reconnection line passes the stagnation point it would be blown off by the solar wind. Maezawa (1976) has suggested that a stagnation point can be formed in the dayside polar cusp region if there is trapped plasma in that region. In Figure 46 the structure of the northern cusp is illustrated schematically for the case when the reconnection line passing the magnetopause nose is not operating. Bounded by the solar wind and open and closed field lines of the magnetosphere, plasma may be trapped in the cusp region shown by the dotted area. The viscous drag by the solar wind would set this plasma in motion as indicated by arrows, and stagnation points would be formed at points *A* and *B*. If the reconnection line passes the point *A*, it can be expected that this reconnection line would be maintained in the cusp region without being carried away by the solar wind.

In passing, let us see what remains in the polar cap in circumstances when the IMF-dependent disturbances are expected to be absent. Figure 47 shows the average distribution of the equivalent overhead current

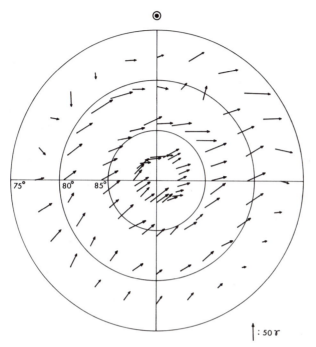

Fig. 47. Equivalent overhead current distribution when IMF is in the range of $|B_y| < 2\gamma$ *and* $0 < B_z < 2\gamma$ (Maezawa, 1976)

vectors for intervals of $|B_y| < 2\gamma$ and $0 < B_z < 2\gamma$. The field vectors are measured from the average of daily mean values on five quiet days in each month for the period studied. The range of B_z is chosen so as to minimize the effects of both nose-type and cusp-type reconnections. The B_y-dependence has further been eliminated by using the linear regression coefficients. The principal feature of Figure 47 is the dusk-to-dawn current that can be interpreted to be the extension of the Sq current from lower latitudes (cf. Fig. 95). The skewing of the current on the night side would be due to the presence of the field-aligned current originating from the magnetotail. This effect, which was noted also in Figure 24, will be discussed further in the next Chapter. In addition, the current vectors on the day side of the polar cap show kinks at 9 ~ 10 LT and 14 ~ 17 LT ranges between 78° and 80°. Although these kinks may be compared with the current vortices of Figure 43b, there is no intensification of the anti-sunward current flow on the day side of the pole in the present case. It has been suggested that the foregoing disturbance reflects magnetospheric perturbations localized to the vicinity of the magnetopause (Maezawa, 1976), but its precise nature remains to be clarified.

The S_q^p Phenomenon

On the basis of the preceding knowledge on the response of the polar-cap magnetic field to various components of the interplanetary magnetic field, we wish to interpret the nature of the S_q^p field that has been found in the polar cap when the magnetic activity in the auroral zone is low. The derivation of S_q^p was made, before the IMF data became available, by Nagata and Kokubun (1962), to determine the immediate geomagnetic effect of the solar-wind magnetosphere interaction process. They examined the daily variation in high latitudes during five quiet days and found that the equivalent current system of Figure 48a exists superposed on the extension of the Sq current from lower latitudes. The current system consists of two vortices centered roughly in morning and evening meridians, and its intensity shows strong seasonal variations. (The confinement of their current system to above latitude of $60°$ represents an assumption which was necessary because separation of S_q^p from the Sq field becomes difficult in lower latitudes.) The variation of this type has been designated as S_q^p, meaning the quiet daily variation of the polar type. Characteristically, the equivalent current flow of S_q^p does not show significant enhancement at auroral-zone latitudes. This contrasts S_q^p to the substorm field, which, as described in the next Chapter, is characterized by the presence

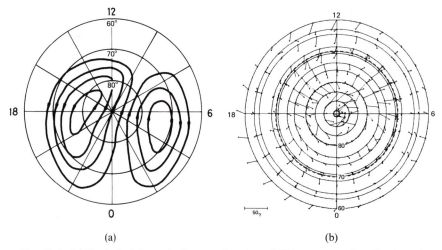

(a)   (b)

Fig. 48a,b. (a) Equivalent ionospheric current system of S_q^p in summer after Nagata and Kokubun (1962). The current between adjacent streamlines is 2×10^4 A. (b) Distribution of horizontal components of the S_q^p field on May 8, 1964 (Kawasaki and Akasofu, 1967). To convert the vectors to equivalent current directions rotate them clockwise by $90°$

of the electrojet flowing along the edge of the polar cap, and makes it likely that DP2 is the principal constituent of their S_q^p field.

The derivation of the S_q^p field has been repeated by several authors by utilizing the data from days that are more quiet than those studied by Nagata and Kokubun (1962), and Kawasaki and Akasofu (1967) have presented Figure 48b as the vector distribution of the real S_q^p; they made a point that Nagata and Kokubun's (1962) S_q^p is not entirely free from contamination by the substorm field. The characteristic difference between Figures 48a and b lies in the direction of the field over the pole; while in Figure 48a the equivalent current over the pole is directed from ~ 22 LT to ~ 10 LT meridian, the vector distribution inside the latitude circle of $\sim 85°$ of Figure 48b is entirely different from what is expected from such a current. This difference makes Kawasaki and Akasofu's S_q^p incompatible not only with Nagata and Kokubun's S_q^p but also with the idea that the driving electric field of S_q^p is directed roughly from dawn to dusk over the pole. Despite this difference, Kawasaki and Akasofu (1973) adopted the interpretation that their S_q^p is associated with the dawn-to-dusk electric field across the polar cap. Thus when it comes to the interpretation of S_q^p, there is no difference between Nagata–Kokubun and Kawasaki–Akasofu. [Indeed, the distribution of geomagnetic vectors calculated by Kawasaki and Akasofu (1973), who adopted the foregoing idea and took into account the effect of the associated field-aligned current, looks more like Nagata–Kokubun's S_q^p than their own.] Akasofu et al. (1973c) suggested that their S_q^p is a permanently existing feature and proposed to interpret DP2 fluctuations as modulations of the S_q^p field. However, subsequent analyses of the polar cap magnetic field by Friis-Christensen and Wilhjelm (1975) and Rostoker et al. (1974) demonstrated that the field distributions in the polar cap during geomagnetically quiet conditions are not always like Kawasaki and Akasofu's observed or postulated S_q^p; this is particularly the case during intervals of northward IMF. Hence it seems more pertinent to consider that DP2 is the basic disturbance field and S_q^p is the apparent daily variation produced by the prolonged or repeated occurrence of DP2 (Nishida and Kokubun, 1971).

III. Implosion in the Magnetotail

III.1 Introduction: Substorm and the Auroral Oval

Undoubtedly, the most lively manifestation of the dynamic nature of the earth's magnetosphere is the auroral break-up. The magnetospheric phenomena that are associated with this explosive release of energy have been the aim of intense research efforts in recent years, and the term magnetospheric substorm has been introduced to designate these phenomena as a whole. The 'polar magnetic substorm' is the manifestation of the magnetospheric substorm in the polar magnetic field. Other aspects of the magnetospheric substorm are similarly called 'auroral substorm,' 'ionospheric substorm,' etc. (Akasofu, 1968). The magnetospheric substorm represents a process in which plasma is accelerated and electric current is generated at the expense of the magnetic field energy stored in the magnetotail.

In the phenomenology of substorms, the auroral oval serves as the basic frame of reference. The auroral oval was originally defined as a region of high occurrence frequency of the aurora as seen in all-sky photographs. Figure 49a is the contour map of the occurrence frequency of the zenith aurora at 24 stations in the northern hemisphere; the frequency is estimated by counting the relative number of half-hourly intervals in which the aurora is seen in the zenith. The auroral oval is a continuous belt that encircles the geomagnetic pole with its center displaced appreciably toward the dark hemisphere (Feldstein, 1963). Hence a station rotating with the earth observes highest frequency of the auroral occurrence at different local times depending upon the geomagnetic latitude of the station. (The latitude range where the aurora appears most frequently around midnight is called auroral zone; it is centered around 67°.) The idea that the auroral oval is the seat of the instantaneous occurrence of the aurora received substantial support when it became possible to command the overall view of the polar region by the scanning photometer carried on board the polar-orbiting satellites. Figure 49b is an example of the scanning photometer data taken at the wavelength of 557.7 nm. In this example discrete auroral arcs are seen at a wide range of local times,

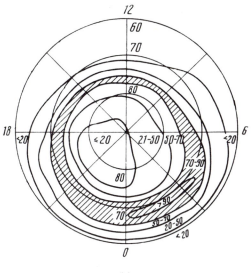

(a)

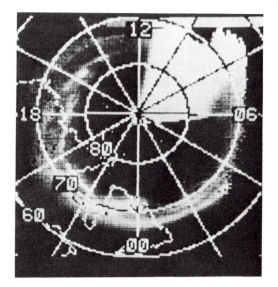

(b)

Fig. 49a,b. (a) Auroral oval statistically defined (Feldstein, 1963), and (b) the snapshot (taken in 15 min) of the auroral oval (Lui et al., 1975a) and their distribution is clearly oval-shaped, the corrected geomagnetic latitudes of the arcs being $\sim 70°$, $\sim 75°$, and $\sim 76°$ in the midnight, dusk, and midday sectors, respectively (Lui et al., 1975a). In addition, a diffuse aurora is seen to envelop the arcs in most parts.

In terms of the open model of the magnetosphere the auroral oval is considered to follow the boundary of the open field lines defining the polar cap. The aurora on the day side of the oval is produced by particles that have the spectral characteristics of the magnetosheath plasma (Heikkila et al., 1972). These particles are most likely to have found access to ionospheric heights via open field lines (cf. Fig. 40). Particles responsible for exciting discrete arcs on the night side of the oval, on the other hand, originate from outer layers of the plasma sheet (Winningham et al., 1975), which probably is bounded by the field lines that pass the neutral line in the distant magnetotail (Feldstein, 1973). As described in the previous Chapter, the latitude of the polar cusp decreases with an increase in the southward component of IMF in association with the erosion of the day-side magnetosphere (Burch, 1972b), and latitudes of the auroral oval in the dusk and midnight sectors have also been found to decrease when the IMF southward component increases (Pike et al., 1974). Thus the flux of open field lines is increased and the flux content of the magnetotail is enhanced under the influence of the southward component of IMF. This brings the magnetotail to an unstable state, and eventually a process is initiated that disposes of the excess flux of open field lines.

The idea that the magnetospheric substorm is the consequence of an impulsive enhancement of the reconnection rate in the magnetotail was put forth in the mid-1960s by Atkinson (1966) and others, and working models of the substorm have since been developed on the basis of this idea. This Chapter deals with the magnetospheric substorm with emphasis on its geomagnetic aspect. Following the morphology of the magnetic substorm, we examine the associated features in the magnetotail and try to present a physical picture of the magnetospheric substorm as the manifestation of the unsteady nature of the magnetospheric convection. Since the energy liberated in the magnetotail is brought to the ionosphere in the form of electric current and precipitating particles, the coupling between the ionosphere and the magnetosphere must also be examined. The discussion of the particle aspect of the substorm, however, is kept to a minimum, because the subject will be developed in another monograph of this series.

III.2 Magnetic Substorm

Ground observations of aurora and magnetic field have shown that the development of a substorm tends to follow a certain prescribed course.

Central to that development is the expansion phase where the aurora breaks up and the magnetogram records show perturbations whose trace has been compared to bays found in geographical maps.

Definitions of Expansion Phase

Akasofu (1968) divided the auroral substorm into two phases: the expansive phase and the recovery phase. According to him the first indication of the expansive phase is a sudden brightening of one of the quiet arcs lying in the midnight sector of the oval or a sudden formation of an arc. The subsequent development of the auroral substorm was summarized in a schematic diagram that has since been widely quoted, and in Figure 50 we have reproduced a series of satellite photographs obtained during different substorm events but arranged to represent the sequence of the development found by Akasofu. If these photographs had been taken during the same substorm event, the interval between successive photographs would be a few minutes (Akasofu, 1975). In most cases, the brightening of an arc or the formation of an arc (which signifies the onset of the expansive phase) is followed by its rapid poleward motion, resulting in an 'auroral bulge' around the midnight sector. As the auroral substorm progresses, the bulge expands in all directions; in the evening side of the expanding bulge the 'westward traveling surge' travels rapidly along the oval, and in the morning side of the bulge the arcs disintegrate into 'patches.' The expansive phase ends and the recovery phase starts when the expanded bulge begins to contract. The duration of the expansive phase is 30 min to 1 h (Akasofu, 1968).

(In the following, we shall use the term expansion phase instead of expansive phase so as to be consistent with the usage of the word in related terms such as recovery phase and growth phase to be introduced later.)

Although the photographs taken from satellites have added powerful means to the study of the overall structure of the aurora, the temporal resolution that is currently attainable is not yet sufficient for following the development of a single substorm. Hence the dynamic development of an auroral substorm has to be described in terms of the ground observations by all-sky cameras and meridian-scanning photometers. The top panel of Figure 51a shows the temporal variation of the auroral latitude observed by a meridian chain of auroral observatories. The observation was made in the Alaskan meridian where the geomagnetic midnight is around 11 UT. The heavy solid lines represent the position and continuity of auroral forms, and the shaded area represents the latitudinal extent of the auroral oval. Feldstein and Starkov's (1967) statistical oval

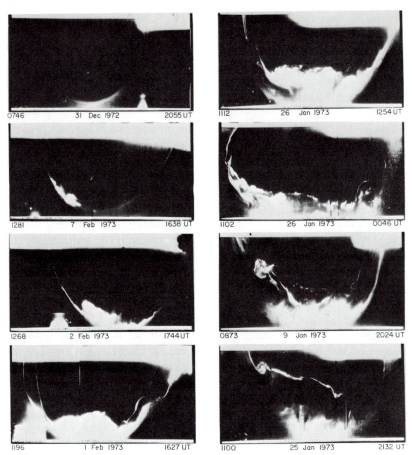

Fig. 50. Development of the aurora during the expansion phase. The midnight meridian is toward bottom center of each panel (Akasofu, 1975)

corresponding to the geomagnetic Q index[8] of 3 is also shown for reference. Superposed on the diurnal trend, which is in gross agreement with the statistical oval, there are rapid meridional motions. In particular, at ~ 1040 UT there occurred rapid poleward motion and equatorward spread that signified the onset of the expansion phase. This was preceded by a pronounced equatorward motion that lasted for ~ 60 min. The same features were observed also at ~ 1130 UT (Snyder and Akasofu, 1972).

---

[8] The Q index is a quarter hourly measure, on a quasi-logarithmic scale, of the magnetic activity for stations at latitudes higher than 58°. For details, see Bartels (1957).

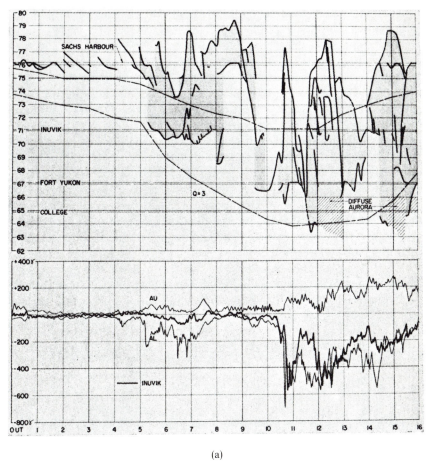

(a)

Fig. 51a. A cross section of the auroral oval during a 16-h interval (top) and the corresponding auroral-zone magnetic data (bottom) (Snyder and Akasofu, 1972). Latitudes of stations are given in the corrected geomagnetic coordinates

The bottom panel of Figure 51a shows the magnetic field data from auroral-zone stations in the form of the AU and AL indices[9]. The

---

[9] Indices AU, AL and AE representing the magnetic activity in the auroral zone are defined in the following way. First, disturbance fields in the H-component are measured at several observatories in the auroral zone relative to levels during quiet intervals. (Efforts are made to use observatories which are equally spaced in the best possible way.) All these data are plotted together, and the upper and the lower envelopes of the superposed plot are defined to be AU and AL. AE is the separation between them; namely $AE = AU - AL$ (Davis and Sugiura, 1966).

H-component magnetogram of the Inuvik station (geomag. lat. 70°), which is located at the center of the meridian chain, is also shown. It is seen that the expansion phase onset at ∼1040 marks a sharp decrease in AL, which, as seen by the Inuvik magnetogram, represents the occurrence of a 'negative bay' in the midnight sector. AU is also enhanced at this time, reflecting the occurrence of a 'positive bay' in the evening sector. The expansion-phase onset at ∼1130, however, is not clearly marked in these magnetic field data. The left panel of Figure 51b shows a set of magnetograms from low-latitude stations for the corresponding interval; these seven stations are selected so as to encircle the earth in the best possible manner. It is seen that at the expansion-phase onset of ∼1040, the *H* component started to increase around midnight (at Honolulu) and to decrease in the evening-to-afternoon sector (at Memambetsu, Irkutsk, and Tbilisi), as indicated by the second of the two arrows in the Figure. The first of these two arrows shows that another change having the same character as the foregoing occurred about 20 min earlier, and this seems to correspond with the occurrence of a smaller negative bay at Inuvik at the corresponding moment. The likely counterpart of the ∼1130 expansion phase is the increase in *H* observed at Memambetsu and Guam about 15 min earlier than 1130.

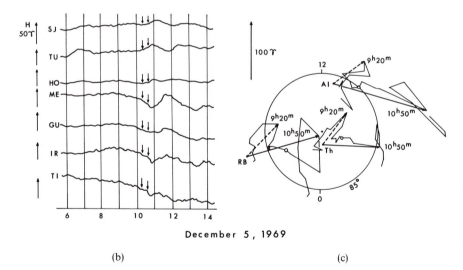

December 5, 1969

(b)   (c)

Fig. 51b,c. (b) Low-latitude magnetograms corresponding to part of (a), and (c) loci of magnetic disturbance vectors at three polar stations. Small circles show the timing of the onset of the expansion phase. Geographic longitudes of the stations used in (b) are, from the top, San Juan (295°), Tucson (249°), Honolulu (202°), Memambetsu (144°), Guam (145°), Irkutsk (104°) and Tbilisi (45°). (Kokubun and Iijima, 1975)

The preceding example illustrates the difficulty of identifying the onset of the expansion phase in a unique and unambiguous manner. Among the three cases of onsets mentioned, only one, namely, the ~1040 case, is clearly demonstrated by both auroral and magnetic observations. The ~1020 onset, which is found from magnetic observations, was not noted in the auroral analysis by Snyder and Akasofu, and the ~1130 onset, which is clear in the auroral data, appears in magnetograms as a weak signal in a spatially very limited region, and it can easily be missed. Technical difficulties should certainly be among the causes of these discrepancies; the brightening of the arc may be missed if it is not strong enough compared with the background, and the magnetic signature of an expansion phase may be masked if other disturbances of comparable magnitudes are already in progress. Nevertheless, at the time of writing this, it does not seem to be a proven thesis that there is a unique one-to-one correspondence between occurrences of expansion phases of auroral and magnetic substorms. Moreover, recent analysis of the satellite auroral photographs has revealed that auroral substorms can be divided into two types, confined and widespread, on the basis of their longitudinal extent, which appears to be related to their intensity (Lui et al., 1975b). It remains to be seen whether expansion phases of confined and widespread auroral substorms are associated with identical global signatures in the magnetic field.

Hence at present it seems inevitable to choose either auroral or magnetic signature in order to define the expansion phase in a consistent way. Although substorm phases were originally defined by the auroral data, it is the magnetic data that are continuously obtainable regardless of weather and lighting conditions. The separation of the global signatures from the localized ones can be made reasonably well by using the magnetic data from the existing network of magnetic observatories, but it is rather difficult to achieve the same by the very limited number of all-sky camera pictures that are routinely produced. For these reasons, it seems more practical to refer to the magnetic data to monitor the onset of the expansion phase.

Development of Magnetic Substorm

The classical magnetic signature of the expansion-phase onset is the onset of a sharp negative bay in the auroral zone. This feature, which was demonstrated by the Inuvik magnetogram and the AL index in Figure 51a, is illustrated by two more examples in Figure 52 using the AL index. The corresponding low-latitude observations are reproduced in the bottom panel by the superposition of magnetogram traces (H components)

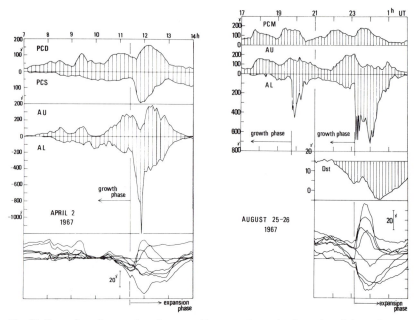

Fig. 52. Examples of magnetic substorms. Data are shown in the order of the polar-cap magnetic indices (PCD, PCS, or PCM), the auroral-zone magnetic indices (AU and AL), the low-latitude magnetic index (Dst, for the right-hand case only), and the superposition of low-latitude magnetograms (Iijima and Nagata, 1972)

from several representative stations. It is seen that the *H* component in low latitudes changes distinctly at the onset of the expansion phase; the increase or decrease in *H* observed here has been called low-latitude positive bay or low-latitude negative bay. The onset time of these low-latitude bays usually differs slightly from place to place, because the disturbed region tends to spread and/or shift with the progress of the event, but when the earliest onset of the low-latitude positive bay is identified, its timing provides a very useful measure of the onset time of the expansion phase (Hones et al., 1971a; Iijima and Nagata, 1972; McPherron, 1972). The obvious advantage is that since it is monitored at large distances from the auroral oval it is less liable to be affected by localized fine structures, and the global development of substorms can be expected to stand out. The earliest onset of the low-latitude positive bay tends to take place slightly before midnight, and the magnitude of the low-latitude positive bay in the night sector is proportional to the intensity of the westward electrojet, which flows along the auroral oval and gives rise to the auroral-zone negative bay, integrated over its entire latitudinal width (Kamide and Akasofu, 1974).

(There is one more feature that has proved useful as a signature of the onset of the expansion phase: onset of micropulsations of the pi2 type. The discussion of this topic, however, is deferred to Chapter V.)

The global distribution of the disturbance field during the expansion phase is shown in the right panel of Figure 53 by an equivalent current system. The characteristic feature is the strong concentration of the equivalent current to the auroral oval (called auroral electrojet) and the connection of the polar-cap and the low-latitude equivalent currents to this electrojet. In the polar cap the equivalent current comprises two vortices, of which one is centered around midnight and the other in the afternoon sector. The former vortex originates from the westward auroral electrojet associated with the (high-latitude) negative bay, while the latter vortex is from the eastward electrojet associated with the (high-latitude) positive bay. During the expansion phase (and also the recovery phase) these current vortices show variations both in intensity and configuration, and the temporal behavior of the two vortices is usually not coherent (Kamide and Fukushima, 1972). High-latitude negative bays that occur around midnight are usually sharper and deeper than those that appear in the dawn sector.

In the premidnight sector, labeled zone of confusion in the right panel of Figure 53, the westward and the eastward electrojets tend to overlay, the eastward electrojet extending toward midnight on the equatorward side of the westward electrojet. The interface between the two electrojets

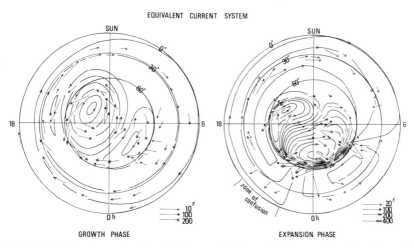

EQUIVALENT CURRENT SYSTEM

GROWTH PHASE   EXPANSION PHASE

Fig. 53. Equivalent current systems for the growth phase and the expansion phase. Current density between adjacent streamlines is 3×10^4 A and 4×10^4 A for the growth and the expansion phase (Iijima and Nagata, 1972)

is called the Harang discontinuity (Sugiura and Heppner, 1965). Latitude profiles of the magnetic perturbation that is recorded below that structure are presented in Figure 54. Positive and negative peaks in the H component reflect the presence of the eastward and westward electrojets; latitudes of their centers can be deduced from the zero-crossing of the Z component to be roughly 63° and 70°, respectively (Rostoker and Kisabeth, 1973). Note, however, that there are occasions where the high-latitude positive bay, and hence the evening-side polar current vortex, appears to be missing, while the night-side vortex always exists as a characteristic feature of the expansion phase.

Since the expansion phase represents a stage in which energy is supplied explosively to the ionosphere, it should be preceded by an interval during which the rate of energy build-up exceeds the rate of its release somewhere in the magnetosphere. In recent years the problem of whether the process of the energy build-up can be monitored by the ground magnetic data as a disturbance that is distinguishable from the expansion phase has been a subject of much concern. Iijima and Nagata (1968) reported that for 1 ~ 2 hours before the onset of the expansion phase there appears in the polar region a mode of disturbance that is expressible by two vortices of the equivalent ionospheric current (Fig. 53, left panel). McPherron (1970) also found the corresponding feature in the auroral-zone magnetic data and proposed that it signifies the 'growth phase' of a substorm. What appears to be the electric counterpart of this feature has been detected in the auroral zone near local midnight as the presence

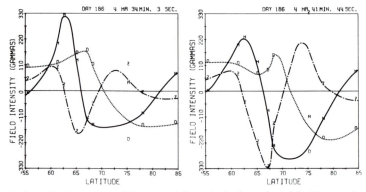

Fig. 54. Latitudinal structure of the auroral electrojet in the region of the Harang discontinuity. The observation was made at the Alberta chain located at the corrected geomagnetic longitude of around 300°E. Traces Z, H, D represent latitudinal profiles of the disturbance field in each component. Two sections at successive moments are shown (Rostoker and Kisabeth, 1973)

of the westward electric field prior to the onset of the expansion phase (Mozer, 1971).

The equivalent currents of geomagnetic disturbances that are suggested to signify the growth phase and the expansion phase bear certain similarities when attention is confined to their auroral-zone portions; both currents correspond to an increase in the H component in the evening sector and a decrease in it in the morning sector. This led Akasofu and Snyder (1972) to suggest that the proposed growth-phase signatures were in fact due to expansion phases which happened to occur when the auroral oval was contracted and located far poleward of the conventional auroral zone, or to the arrival of westward-traveling surges originating from the expansion initiated far eastward of the observation site.

However, disturbance fields during the growth phase and the expansion phase have distinct differences in the central polar cap and in the near-midnight region of the low-latitude zone. In the polar cap the difference lies in the direction of the equivalent current vectors as shown in Figure 51c for the ~ 1020 and ~ 1040 December 5, 1969 events discussed earlier. This Figure shows the end points of the equivalent current vectors (measured from the extremely quiet day level) during a ~ 4-h interval encompassing the expansion phase onset. Vectors are produced every 10 min and data from three polar stations (Alert, Thule, and Resolute Bay) are shown together. It is seen that at all three stations the direction of current vectors deflects toward earlier local times around the onset of the expansion phase as expected from the equivalent current systems of Figure 53 (Kokubun and Iijima, 1975). The difference in signatures in low latitudes is that during the growth phase the H component is not yet increased anywhere, including the premidnight-to-morning sector where the low-latitude positive bay is to be observed during the expansion phase; as shown in the bottom panel of Figure 52, the low-latitude H tends instead to decrease during the growth phase in a broad sector. Thus the concept of the growth phase as the precursive stage of the expansion phase seems to have sound observational basis in the ground magnetic data at least for some substorm events. (In the top panel of Figure 52 the development of the growth phase in the polar cap is expressed by using PC indices defined by the authors.)

In the polar cap the equivalent current system of the growth-phase disturbance is essentially the same as that of DP2. Difference is noted, however, in low latitudes; during the growth phase the H component is decreased in the afternoon–evening sector in association with the intensification of the night-to-day flow of the equivalent current across the polar cap, but in the same sector the DP2 field that is coherent with the enhancement of the polar-cap night-to-day current is the increase in the

H component. It thus appears that the geomagnetic disturbance observed during the growth phase is the sum of DP2 and one more constituent that acts to produce the southward disturbance field in low-latitudes (Iijima and Nagata, 1972). Because the energy input to the magnetotail is probably not inhibited by the onset of the expansion phase, the disturbance field that stands out during the growth phase may well continue through the expansion phase; indeed DP2 fluctuations have been found to coexist with the intense auroral electrojet and with the auroral activity (Nishida, 1971a; Kawasaki and Akasofu, 1972). Suggestions have been made that the broad high-latitude bays that appear in morning and evening sectors are due largely to the disturbance of the DP2 mode with intensifications along the auroral oval (Troshichev et al., 1974). The intensification is attributable to the conductivity enhancement due to the ionizing effect of precipitating particles.

In the magnetic substorm studied in Figure 51, onsets of expansion phases are observed twice at ~ 1020 and ~ 1040 in close succession. Cases having multiple onsets like the preceding are frequently encountered. An example in which four onsets are recognized is reproduced in Figure 55, where left and right panels contain low- and high-latitude records of the *H* component. (Low-latitude records are shown in the order of decreasing local time from 24 to 17 LT.) The onsets A and B can be identified by the occurrences of low-latitude positive bays. Although low-latitude positive bays at A and B are observed in different local-time sectors within ~ 6 min, they cannot be interpreted to represent a continuous longitudinal shift of a single disturbance. Note that onsets of positive bays A and B are simultaneous where they are observed. The corresponding sharp negative bays in the auroral zone are observed also at stations slightly separated longitudinally (Great Whale and Churchill). The other two onsets C and D are less well-defined in low-latitude records but are identifiable in auroral-zone magnetogram records. Thus the expansion phase sometimes starts in multiple steps and the disturbance observed during each step can have different spatial structures on the ground (e.g., Wiens and Rostoker, 1975). Although the center of the westward electrojet activity shifted successively northwest in the preceding example, cases have also been found where the activity moved eastward from the meridian of the initial onset.

The multiplicity of the expansion-phase onset might raise a question of whether a reinforcement of the energy build-up can be detected in the ground field before the *n*th onset when $n \geq 2$. This question is very hard to answer, however, because DP1, namely, the disturbance associated with the intensification of the auroral electrojet, tends to be predominant in such intervals. [In this connection, it is noted that the definitions of

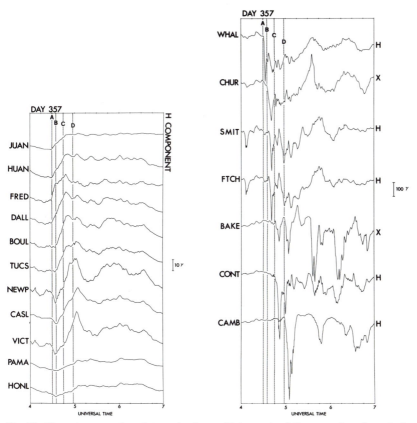

Fig. 55. Magnetograms of a substorm having multiple onsets of the expansion phase. *Left*: low-latitude data, *right*: high-latitude data (Wiens and Rostoker, 1975)

the growth phase employed in critical studies by Wiens and Rostoker (1975) and Kamide et al. (1975) are different from the one adopted here. Wiens and Rostoker took the depression in the low-latitude H component as the growth-phase signature, but in the evening sector this feature occurs during our 'expansion phase' also (cf. Fig. 52), and we interpret the weak substorm studied by Kamide et al. as the first onset of a multiple expansion event.] Similarly, during intervals of continuing high activity DP1 tends to dominate the disturbance field, and the growth-phase signatures are usually difficult to separate from the overall disturbance field in the ground magnetic field.

The transition from the expansion phase to the recovery phase is to be considered to occur when the westward electrojet around midnight reaches the highest latitude and attains the maximum width, according to Akasofu's original definition of substorm phases in terms of auroral features (Kisabeth and Rostoker, 1974). More conventionally, however, the times when the auroral-zone negative bay around midnight becomes deepest or when the low-latitude positive bay reaches its peak have frequently been taken to mark the end of the expansion phase. Although these different signatures are not entirely simultaneous, they have been found to agree roughly with the time when an important structural change takes place in the magnetotail. This topic will be extended in the next Section.

In light of the present understanding of the development of the magnetic substorm, it is tempting to interpret the equatorward movement of the auroral arc as noted in Figure 51a as a manifestation of the growth phase in the auroral substorm. Such an interpretation has indeed been proposed (e.g., Pudovkin et al., 1968). Although alternative causes like the diurnal shift of the auroral oval (Snyder and Akasofu, 1972) or the after-effect of the previous substorm have also been suggested as the cause of this movement, the speed of the movement is much faster than what is expected from Feldstein's oval pattern, and there are instances where the equatorward movement was shown to precede the onset of the expansion phase by about 1 h. In these instances the equatorward movement started 15 ~ 25 min after the southward turning of IMF (Pike et al., 1974).

Solar-Wind Effect on Substorm Activity

It was noted in the last Chapter that indices representing the magnetic activity in the auroral zone show positive correlations with the southward component of IMF. Although disturbance fields of various types can contribute to these indices, the principal constituents are DP1 and DP2, and high values of indices usually arise from the intense DP1 field. Hence it has been concluded that the auroral electrojet is made more intense when the southward IMF component is larger. As a matter of fact available indices like AE or Kp[10] are not very accurate measures

---

[10] Kp index is a quasi-logarithmic, 3-hour index (0 through 9) of the high latitude magnetic activity which is based on magnetic observations at 13 (mostly northern) stations located between 46° and 63° latitude. The basic material is the K index derived at each observatory from the range of variation (in each 3-hour interval) of each component H, D, and Z (or, X, Y, and Z). The detailed process for the calculation of the Kp index is described in Rostoker (1972). The ap index is the transformation of Kp to a linear scale, and the Ap index is the average of eight ap values for a given day. (The Am index introduced earlier is a modified version of Kp where activities in the northern and the southern hemispheres are more equally represented and corrections are made for non-uniform distributions of the stations.)

of the electrojet intensity when IMF is directed northward, because under such circumstances the electrojet tends to flow along the contracted oval, which is far away from the auroral-zone stations whose records are used for constructing the indices (Akasofu et al., 1973b). Nevertheless the preceding conclusion was substantiated when the total intensity of the electrojet, integrated across the latitudinal width, was derived from magnetic observations on a meridian chain of stations (extending from geomagnetic latitudes of $\sim 60°$ to $\sim 80°$) and compared with the IMF data (Kamide and Akasofu, 1974). Furthermore, a correlative study between IMF and the occurrence of auroral break-ups has confirmed that the occurrence frequency of substorms increases as the southward IMF component increases, although substorms are not entirely absent during intervals of northward IMF (Kamide et al., 1977).

As for the response time of DP1 to the southward IMF component, a value of $1 \sim 2$ h has been derived from the study of those expansion-phase onsets that follow a sharp southward turning of IMF after a prolonged interval of northward IMF (Kokubun, 1971). In such cases the growth phase signatures are observed in the meantime. The delay time of ~ 40 min obtained by the cross-correlational analyses between B_z of IMF and AE (Foster et al., 1971) is intermediate between response times of DP1 and DP2, as expected.

The magnetic activity in the auroral zone shows dependence also on the velocity v_{sw} of the solar wind. This dependence does not appear to be of primary importance, however, since variations of v_{sw} lag behind those of Kp according to results of cross-correlational analyses (e.g., Ballif et al., 1969). The lag of v_{sw} variations behind Kp variations probably reflects the tendency that in the solar wind the maximum of v_{sw} occurs after the maximum of the IMF strength which has a stronger influence on the geomagnetic activity. Nevertheless, auroral-zone magnetic activity is not entirely without an intrinsic dependence on v_{sw}. After the overall dependence on IMF has been eliminated, it has been found that AL is proportional to v_{sw}^2 (Murayama and Hakamada, 1975). AU, on the other hand, is proportional to v_{sw} (Maezawa and Nishida, 1978). The dependence on first power of v_{sw} is attributable to the proportionality of interplanetary electric field on v_{sw}, but the reason for the dependence on v_{sw}^2 remains to be found.

A small fraction of the expansion-phase onsets is triggered by positive sudden impulses (Kawasaki et al., 1971). The triggering of the expansion phase occurs when the magnetosphere is already in the disturbed condition (as expressed by AE $> 100\gamma$) and/or when the north–south component has been negative or decreasing over a period of 30 min when the magnetosphere is suddenly compressed (e.g., Burch, 1972a). An

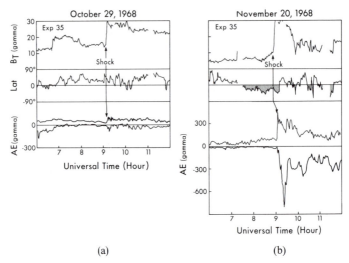

Fig. 56a,b. Interplanetary shock waves leading to an increase in AE (case b) and not (case a). B_T and Lat are magnitude and latitudinal angle of the interplanetary magnetic field (Kokubun et al., 1977)

example of the SI-triggered substorm expansion phase is given in Figure 56b. Characteristically the solar magnetospheric latitude angle of IMF was negative for a few hours before the onset of the expansion phase was triggered by an interplanetary shock wave. In contrast, the arrival of an interplanetary shock wave did not cause the expansion phase in the example of Figure 56a, where the IMF latitude was positive during a few hours' interval that preceded. This suggests that the expansion phase can be initiated by a sudden compression of the magnetosphere only when the loading of the solar-wind energy to the magnetosphere has been proceeding for some time ($\gtrsim 30$ min) on the day-side magnetopause through the reconnection process. Needless to say, majority of expansion phases is started without any triggering effect by the solar wind. The process of the expansion-phase onset should therefore be basically internal to the magnetosphere.

III.3 Substorm in the Magnetotail

In this section we shall see what corresponding phenomena take place in the magnetotail during magnetic substorms. The principal point of interest is to survey the structural change of the tail that occurs at the onset of the expansion phase and thereby to deduce the mechanism that causes the substorm phenomenon. During intervals that precede the expansion-phase onsets the build-up of energy should also be discernible in the magnetotail.

Representative Case Studies

The magnetotail consists of the lobe and the plasma sheet (c.f. Figure 5). In the lobe that constitutes the higher latitude part of the tail, the magnetic field energy far exceeds the plasma kinetic energy. In the plasma sheet that occupies the lower latitude portion, the plasma kinetic energy dominates. Hence it is the plasma sheet that is the seat of dynamical processes, while it is mainly the lobe that acts as the reservoir of the magnetic field energy. Let us look at the observations in the tail lobe first.

Magnetic field data obtained in the lobe are compared with ground magnetograms in Figure 57. On this occasion two satellites (IMP-C and AIMP-D) were simultaneously orbiting the magnetotail allowing us to compare substorm behaviors in the central and the duskward regions of the tail lobe. The second and third panels of this Figure display ground magnetograms from low-latitude [Guam (4°) and Tashkent (32°)] and auroral-zone [College (65°) and Dixon (63°)] stations, and vertical lines indicate expansion phase onsets determined by low-latitude positive bays. The second event is a double-onset event. The top panel of this figure shows the solar-magnetospheric x component B_x of the magnetic field observed by each satellite, whose positions at the beginning and the end of the interval are written within the figure by the solar magnetospheric coordinates. The figure clearly shows that both satellites recorded a peak in B_x at (or closely behind) each expansion-phase onset. Since the magnetic field is nearly parallel to the x-axis in the lobe of the tail, the observation means that the field magnitude in the tail lobe decreased at each onset of the expansion phase. These decreases were preceded by field increases, which can be regarded as signifying the growth phase of substorms in the magnetotail.

Such behavior of the lobe field has been recognized on numerous occasions at $15R_E \lesssim |x| \lesssim 80R_E$ (Fairfield and Ness, 1970; Camidge and Rostoker, 1970). As exemplified above the decrease in the magnetic field at the onset of the expansion phase is observed to begin nearly simultaneously (within several minutes) in a wide longitudinal sector of the tail lobe (encompassing at least $|y_{SM}| \lesssim 10R_E$). It is noteworthy that the double onset of the expansion phase corresponds to a double peak in B_x.

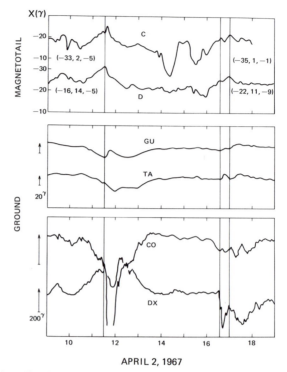

Fig. 57. Dual satellite observations of substorms in the tail lobe and corresponding ground magnetic data (Nishida and Nagayama, 1975)

The above sequence of change in the magnitude of the field in the lobe is accompanied by a change in the diameter of the tail. This is demonstrated in Figure 58 which displays magnetic field records obtained during substorms when a satellite happened to be located in the neighborhood of the tail magnetopause. Six examples are shown together in the left-hand column aligned by the onset time of the expansion phase. (Low-latitude magnetograms used to identify the expansion-phase onsets are reproduced in the right-hand column.) Only the observations made inside the magnetosphere are reproduced, the records being bounded by vertical straight lines at the magnetopause crossings of the satellite. The feature to be noted is that the satellite tends to enter the magnetotail prior to the onset of the expansion phase and leave it after the onset. Since essentially the same feature was observed regardless of whether the satellite was inbound or outbound, it appears that the magnetotail increases its diameter during the growth phase and shrinks when the expansion phase sets in. Inside the magnetotail the satellite observed

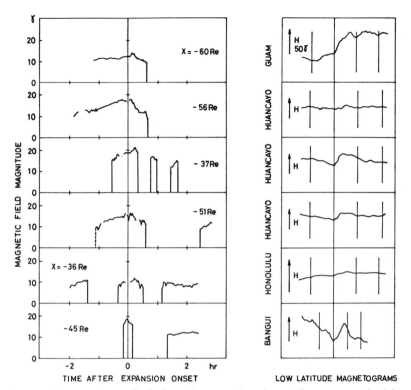

Fig. 58. Tail magnetic records during substorms when the satellite was near the lobe-magnetosheath boundary. For reference, low-latitude magnetograms are reproduced in the right-hand column (Maezawa, 1975)

magnetic field variations typical of the tail lobe, namely an increase during the growth phase and a decrease starting roughly at the onset of the expansion phase. Combining the observations on the tail diameter and on the lobe-field magnitude, we are led to conclude that the flux content of the magnetotail builds up during the growth phase and is relieved during the expansion phase. The increment ΔF_T of the tail flux at the peak of the growth phase (just before the onset of the expansion phase) is estimated to be $1 \sim 3 \times 10^8$ Wb, which amounts to a $10 \sim 30\%$ increase in the flux contained in each lobe of the tail (Maezawa, 1975).

In the low-latitude region, on the other hand, substorm features seem to depend substantially on whether the observing site was inside or outside the radial distance of about $15R_E$. An example of the low-latitude observation made inside this distance is shown in Figure 59. It is a case whose expansion-phase onset (0714 UT according to low-latitude positive bay data) is indicated by the second of the dashed lines. The observation was made around the predicted position of the tail mid-plane, and near the midnight meridian, at the radial distance of about $8R_E$. The characteristic feature that was observed in this case was the sudden increase in the northward component B_z of the field at the onset of the expansion phase (see fourth panel). When combined with the observation of other field components, this B_z increase was found to signify the change of the field orientation to the one that was more dipole-like. In the interval that preceded, B_z was much lower than the level expected for the local dipole field. In particular, for $20 \sim 30$ min immediately before the expansion-phase onset the field line was stretched further. Thus further extension of field lines and the sudden recovery to a dipole-like configuration corresponded to the growth and the expansion phase, respectively.

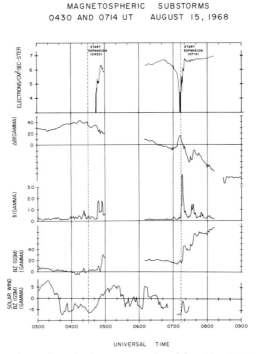

Fig. 59. Substorm observations in the magnetotail at $|x| \sim 11R_E$ (0430 event) and $8R_E$ (0714 event). From the top, (1) flux of electrons with energy > 50 keV, (2) measured field minus dipole field magnitude, (3) variability of the field, (4) solar magnetospheric z component of the field, (5) the IMF's solar magnetospheric z component (McPherron, 1972; McPherron et al., 1973a and b)

The top panel of Figure 59 shows the particle data. It shows the flux of > 50 keV electrons observed simultaneously, and the reduction of the flux and the sudden recovery of the flux were the features observed during the growth phase and at the expansion-phase onset, respectively. Thus the distribution of plasma changed systematically in association with the configurational change of the magnetic field lines.

(Another expansion-phase onset shown in Figure 59, which is indicated by the first of the dashed lines at 0430 UT, was observed when the satellite was still orbiting the relatively high latitude region and at $|x| \sim 11 R_E$. Hence the plasma was not observed until some time after the onset of the expansion phase, and the observed behavior of the magnetic field strength (see second panel) was of the lobe type in the meantime. A significant increase in B_z that signified the rotation of the field direction toward a dipole-like one was observed in association with the entry of the satellite into plasma.)

The bottom panel of this figure shows the north-south component of IMF. It is seen that IMF had a southward component before the expansion-phase onsets, and the precursory features like the increase in the lobe field strength in the first case and the decrease in the electron flux in the second case are seen to have started roughly when the IMF direction turned from northward to southward.

Field and particle observations made in the low-latitude magnetotail at distances much greater than $8 R_E$ show markedly different behaviors as compared to the 0714 UT case described above. Two examples are reproduced in Figure 60. These were obtained at $|x| \sim 30 R_E$ on the dawn and dusk sides of the magnetotail (at $|y_{SM}| \sim 9 R_E$), respectively. The figure shows, from the top, a ground magnetogram from a low-latitude premidnight station, z- and x-components of the tail magnetic field, count rate of ambient plasma-sheet protons having energy of 2.8 to 4.4 keV, and flux of energetic electrons above 0.5 MeV. A vertical line indicates onset time of the expansion phase as determined by the earliest onset of low-latitude positive bays. The plasma proton flux in the third panel demonstrates that the satellite was inside the plasma sheet at the onset of the expansion phase.

In both events of Figure 60, the z-component of the field sharply decreased (namely, moved southward) soon after the onset of the expansion phase. That is, the north-south polarity of the magnetic field became opposite to that of the earth's dipole field, in the very interval when a more dipolar configuration was resumed in the region much closer to the earth; as seen in the previous example. In the first (March 23) event, the

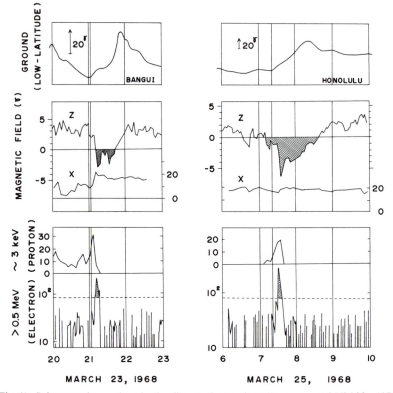

Fig. 60. Substorm observations in the distant plasma sheet (Terasawa and Nishida, 1976)

southward turning of the z-component was associated with an increase in the x-component of the field and a decrease in the plasma proton flux to be background level which indicate that the boundary of the thinning plasma sheet passed the satellite soon after the onset of the expansion phase. Recall that the enhancement of the particle flux was observed in the corresponding interval at $|x| \sim 8R_E$. The flux of energetic electrons rose impulsively at the expansion-phase onset and peaked roughly at the time when the north-south component of the field crossed zero. In the second (March 25) event, both plasma and x-component data show that the observation was made in the boundary region of the plasma sheet, and that the satellite was engulfed in the plasma sheet only for a brief interval around the onset of the expansion phase. The flux of energetic electrons maximized roughly at the time of the sharp southward movement of the z-component (Terasawa and Nishida, 1976).

Energetic protons are also produced during the expansion phase. Unlike energetic electron bursts which do not show pronounced dawn-dusk asymmetry, the occurrence frequency of the energetic proton bursts is strongly biased toward the dusk sector. The majority of intense bursts of 0.3–0.5 MeV protons occurs in the dusk magnetotail (Sarris et al., 1976), and high-intensity bursts of $E_p > 50$ keV protons have been observed only in the dusk sector (Keath et al., 1976).

An anti-earthward flow of the ambient plasma population in the plasma sheet has frequently been observed in the thinned plasma sheet. An example of such an observation is presented in Figure 61. The records were obtained in the plasma sheet at $|x| \sim 30R_E$, and the onset time of the expansion phase was identified to be ~ 0555 UT by the low-latitude magnetogram. The north-south component B_z of the tail field turned southward around 0557 UT and stayed southward most of the time until about 0628 UT (see bottom panel of Figure 61a). The thinning of the plasma sheet is displayed in the bottom panel of Figure 61b as the decrease in the electron flux, but the flux did not entirely drop to the background level. This made it possible to monitor the flow of protons inside the plasma sheet almost throughout the course of the event. The flow velocity vectors displayed in the upper panel of Figure 61b show the occurrence of rapid anti-earthward flows during the interval of the southward B_z. The flow direction was reversed and became earthward when the magnetic field resumed the northward polarity (Hones, 1976; Hones et al., 1974). The simultaneous occurrence of the southward B_z and the anti-earthward flow of the plasma sheet protons during the expansion phase has been detected also at the lunar distance of $|x| \sim 63R_E$ (Burke and Reasoner, 1973). Thus in the distant magnetotail a dynamical reconfiguration is apparently taking place following the expansion-phase onset.

Representative magnetograms from auroral-zone stations corresponding to the above example are reproduced in Figure 61c. Records are shown in the order of the decreasing local time of the station, and M indicates magnetic local midnight. The auroral-zone disturbance that corresponds most closely to the southward turning of the tail field, observed in the plasma sheet at 0557, is the onset of a sharp negative bay in the slightly pre-midnight meridian (namely, at Meanook), and this bay reached its peak shortly (about 4 min) behind the reappearance of the northward field in the plasma sheet at 0628. 0628 is also the time around which the center of the disturbance in the midnight region shifted poleward from Meanook (62°) to Baker Lake (74°). The association of the plasma sheet recovery with the poleward shift of the westward electrojet has often been recognized (Hones et al., 1970; Hones et al., 1973).

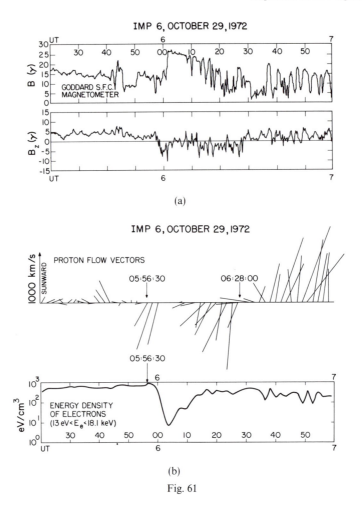

Fig. 61

As exemplified above, substorm signatures observed in the distant magnetotail tend to show a most intimate association with ground magnetic signatures observed in the premidnight to midnight sector. This seems to be the case regardless of the local time position of the satellite, if the satellite is sufficiently far from the earth ($|x| \gtrsim 15 \ R_E$, say) (Nishida and Hones, 1974), and it probably reflects the topology of the mapping of tail field lines to the ground. In contrast, negative bays observed in the morning sector (at Narssarssuaq and Leirvogur in the example of Figure 61c) do not show a distinct signature around 0557 at which a significant structural change occurred in the distant magnetotail. In this example negative bays on the morning side started about an hour before

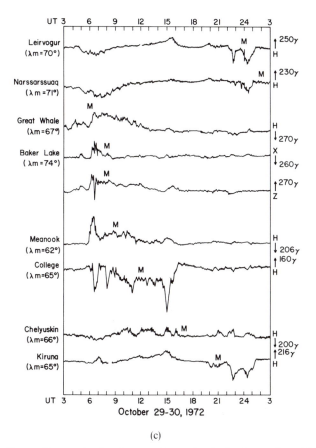

(c)

Fig. 61a–c. Composite records of a substorm observation in the plasma sheet. (a) Magnetic field, (b) proton flow vector and electron energy density, and (c) auroral-zone magnetic data (Hones, 1976)

the expansion-phase onset defined by the low-latitude positive bay (whose earliest onset tends to be found in the premidnight to midnight sector). These morning-side bays are likely to have signified the growth phase of the same substorm.

Indeed there is no *a priori* reason to believe that all the auroral-zone magnetic disturbances (which have been traditionally lumped together under the name of "bays") reflect the same behavior of the magnetotail, and due care is needed in interpreting the nature of the bays. It is thus quite dangerous to depend exclusively on data obtained in the auroral oval and its close neighborhood; a global survey of the magnetic data is often essential for discriminating the nature of the disturbance field.

Phenomenological Summary

A possible way to set the above observations in order is to assume that an X-type neutral line is formed in the near tail region around $|x| \sim 15\, R_E$ during the expansion phase of substorms. The neutral line is supposed to separate the low-latitude magnetotail into two distinct regions; earthward of the neutral line the magnetic field lines resume a relatively dipolar configuration, while on the anti-earthward side the stretching of the field lines continues and the thinning of the plasma sheet proceeds as the plasma flows away. A model based on this idea has been advanced by many (Russell and McPherron, 1973b; Hoffman and Burch, 1973; Hones et al., 1974; Russell, 1974; Terasawa and Nishida, 1976; Hones, 1977), and it is depicted in Figure 62 (Hones, 1977). (Note that the figure is drawn with a much more reduced scale in the direction of the tail axis than in the north-south direction.)

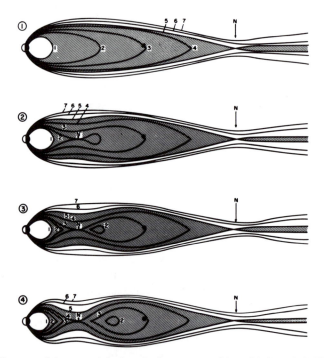

Fig. 62. Progress of the expansion phase in the magnetotail (panel 2 through 8). Five closed field lines (1 through 5) of the pre-expansion-phase plasma sheet are depicted as well as two open field lines (6, 7) that were in the tail lobe before the expansion-phase onset. Fine hatching delineates the plasma of the pre-expansion plasma sheet, while coarse hatching delineates the plasma populating open field lines (Hones, 1977)

In the quiet state (panel 1), the plasma sheet thickness tapers off with an increasing distance but in the 20 $R_E \lesssim |x| \lesssim 60$ R_E range the semi-thickness is roughly 4 R_E near local midnight and about twice as thick near the dawn and dusk edges of the tail. The boundaries are supposedly defined by those field lines that pass the hypothetical neutral line N located at a great distance. This distant neutral line is not thought to take an active part in the substorm process. The magnetic field across the tail midplane is directed northward everywhere when irregular structures are smoothed out.

During the growth phase the thickness is gradually reduced in the near-earth part of the plasma sheet. Although the thinning before the onset of the expansion phase is frequently detected at radial distance r of less than $15R_E$, Vela satellites at $r \sim 18R_E$ do not usually observe the

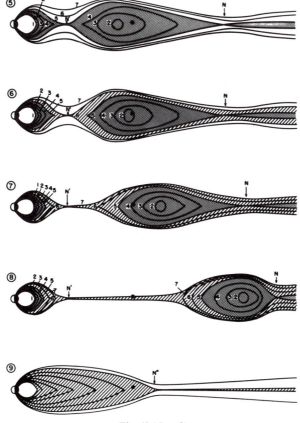

Fig. 62 (*Contd.*)

thinning until the onset of the expansion phase (Hones et al., 1971b), although in some cases the thinning was observed by Vela during the growth phase (Hones et al., 1971a). It thus appears that the spatial extent of the thinning region is rather limited at this stage.

At the onset of the expansion phase (panel 2), an X-type neutral line N' is formed somewhere inside $18R_E$ and the plasma flows away from the neutral line. The formation of the neutral line is due to a sudden weakening (or almost a disappearance) of the cross-tail current in the near-earth part of the plasma sheet. Earthward of this neutral line the plasma sheet expands and the dipole-like field configuration is resumed, while tailward of it a bulge is formed due to a large supply of plasma and field lines from the neutral line N'. In the early stage of the expansion phase the field lines reconnected are those that have been in the plasma sheet (panel 2 through 5). At the neutral line very energetic electrons (extending to the relativistic range) are produced. These electrons spread throughout the field line loops that are produced by the reconnection. The flux of energetic electrons would be the highest at the O-type neutral line located at the center of the loops for the following reasons. First, particles populating the more central of the magnetic loops are those which have passed the X-type neutral line at an earlier stage of the reconnection when the acceleration process may have operated with a higher efficiency. Second, the volume of magnetic loops that are created per unit time by the reconnection is smallest at the onset of the reconnection and increases monotonically as the reconnection progressively involves field lines whose equatorial crossing distance is increasingly further away. The flux would be less diluted and more intense when it occupies a smaller dimension (Terasawa and Nishida, 1976).

As the neutral line N' continues to operate the field lines from the tail lobe begin to reconnect and the content of the open field lines in the magnetotail is reduced (panel 6 through 8). Then the assembly of closed loops, which is no longer tied to the earth, recedes quickly outward through the tail and into the solar wind. This results in an extreme thinning of the plasma sheet (panel 8); the minimum semi-thickness reached at $r \sim 18R_E$ is about $1R_E$ in the midnight sector (Lui et al., 1975c). The thickness of the region of the southward field across the tail midplane is reduced accordingly.

Sometimes the expansion phase onsets are observed in multiple steps. There is an indication that in such cases the structural change that occurred at the earlier expansion-phase onset was relatively limited in the spatial extent, so that the energy that had been brought to the magnetotail could not be released with a sufficiently high rate. Thus the energy storage in the tail was soon resumed, and another structural change that involved a greater volume had to follow, signifying another onset of the expansion phase (Nishida and Nagayama, 1975).

As the substorm starts to recover, the neutral line disappears from the near-tail region and the plasma sheet expands throughout the $|x| <$ $80R_E$ range which has been surveyed. As the earthward flow is observed in this stage at $|x| \sim 30R_E$ it is suggested that the X-type neutral line has shifted to the distant tail and caused the plasma sheet to expand on its earthward side (Hones et al., 1974). Ar $r \sim 18R_E$ the semi-thickness returns to $\sim 5R_E$ (Hones et al., 1973). This might correspond to the poleward shift of the electrojet.

According to the above model, the region of the greatest importance in the substorm process is the low-latitude magnetotail around $|x| \sim 15R_E$ where the X-type neutral line is formed. The conversion of the magnetic field energy to the particle kinetic energy is accomplished at the neutral line as well as at the shock wave extending therefrom. This region of the greatest importance, however, has not yet been surveyed by space probes.

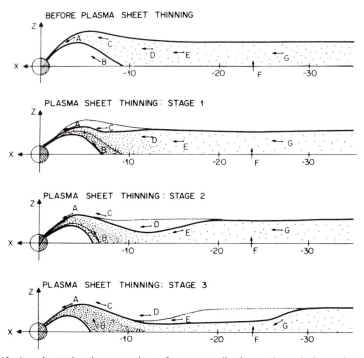

Fig. 63. An alternative interpretation of magnetotail observations during substorms. Magnetic field vectors are depicted at seven representative positions (Lui et al., 1977)

The Vela satellites, which have provided us with a wealth of data on the dynamic behavior of the plasma sheet, traverse the magnetotail at a slightly larger distance. Most other satellites having eccentric orbits are in the lobe when they are around $|x|$ of $15R_E$. Hence it has not yet been possible to establish, by in situ measurements, the acceleration process that is supposed to operate at the neutral line.

In the absence of such measurements, the available data are liable to other interpretations, and Figure 63 shows a model proposed by Lui et al. (1977). In this picture the growth phase is not distinguished, and the thinning of the plasma sheet is thought to propagate continuously from the near-earth region to greater distances down the tail. The southward component of the low-latitude tail field is attributed in this model to the inclination of the edge of the thinning wave that is propagating away; the neutral line is not thought to form in the near-earth region. Since the southward component is observed both at $|x| \sim 30R_E$ and $\sim 60R_E$ and its duration is frequently as long as half an hour or more, however, the wavelength of the thinning wave should be much longer than is depicted in Figure 63. The cause of such substorm signatures as the thinning of the plasma sheet, the antiearthward flow of plasma and the burst of energetic electrons remain to be explained by this model.

III.4 Unsteady Convection in the Magnetosphere

The foregoing phenomenology has been interpreted in terms of the open model of the magnetosphere. In the steady convection model of Figure 64,

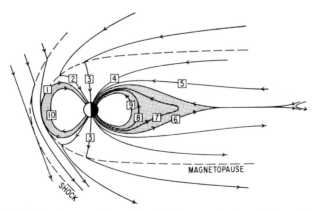

Fig. 64. Idealized steady convection pattern in the magnetosphere. The numbers indicate the successive positions of field lines, with reconnection occurring at positions 1 and 6 (Axford, 1969)

the reconnection at the day-side magnetopause opens the geomagnetic field line at position 1 and convects the open field line tailward to position 6 where it is to be closed by another reconnection process. In order to understand the physics of magnetospheric substorms, however, we have to introduce the concept of unsteady convection.

Build-up of Tail Flux

The presence of the convection would not alter the magnetic configuration of the outer magnetosphere if the magnetic flux that is closed per unit time in the magnetotail were instantaneously equal to the amount of the flux that is opened per unit time along the day-side reconnection line. In that circumstance the flux content of the magnetotail would be kept constant and there would not be an accumulation of magnetic energy in the magnetotail. Observations described in the previous Section have indicated, however, that the tail-flux content tends to increase with time until a rapid reconnection begins to operate in the near-tail region to remove the excess flux from the tail. This means that the reconnection at the distant neutral line, which is supposed to exist in the distant tail where upper (i.e., northern) and lower (i.e., southern) boundaries of the plasma sheet meet, is not efficient enough to balance the day-side re-connection rate. Thus according to our present understanding the magnetospheric substorm is a consequence of the unsteady nature of the magnetospheric convection that is brought about by the inability of the distant neutral line to keep up with the production of open field lines at the day-side reconnection line (Atkinson, 1966; Axford, 1969).

The deformation of the noon–midnight section of the magnetosphere during the growth phase is illustrated in Figure 65. Since the replenishment of closed field lines from the distant tail is not rapid enough to compensate the loss of closed field lines by the day-side reconnection, the day-side magnetosphere erodes as the geomagnetic field lines are successively opened up by the reconnection with IMF. The diameter of the tail, on the other hand, is enlarged.

In terms of electric current the increase in the flux content of the magnetotail is due primarily to intensifications of the line-tying current (see Figure 41) that is associated with the dayside reconnection. The I_L current system acts to reduce the magnetic flux on the dayside of the magnetosphere but to enhance it on the nightside, and consequently to reduce the diameter of the dayside magnetopause but to increase that of the magnetotail. This causes the flaring angle of the tail magnetopause, and hence the angle of attack of the solar wind, to increase. Thus the

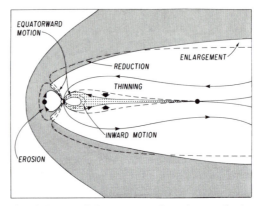

Fig. 65. Configurational change of the magnetosphere that results from the imbalances between the day-side and the magnetotail reconnection rates (McPherron et al., 1973b)

solar wind pressure exerted on the tail is increased, and the flux content of the tail is further augmented as currents flowing on the tail magneto-pause and in the plasma sheet, which becomes compressed, are intensified. The relation between enhancements in the tail flux content and that in diameter and field strength of the magnetotail has been formulated by Coroniti and Kennel (1972b).

The tail is modeled by a bifurcated cylinder of radius $r = R_T(x)$, where x is the distance from the center of the earth along the axis of the cylinder, which is aligned with the solar wind. It is assumed that the flux F_T contained in each lobe is constant with x; namely, the losses of flux due to the flux leakage across the tail magnetopause and also due to the flux closure through the tail midplane are assumed to be negligible as compared to the flux content F_T. Then

$$\varepsilon \left. \frac{B_r}{B_x} \right|_{r=R_T(x)} = \frac{dR_T}{dx} \tag{58}$$

and

$$B_x(x) = \varepsilon \frac{2F_T}{\pi R_T^2(x)} \tag{59}$$

where $\varepsilon = \pm 1$ in the southern and northern lobes of the tail, respectively. The condition of normal stress balance at the tail magnetopause is approximately

$$\frac{B_x(x)^2}{2\mu_0} = Knmv^2 \sin^2 \alpha(x) + P_0 \tag{60}$$

where $\alpha(x)$ is the flaring angle of the tail magnetopause at x, namely, $\alpha = \tan^{-1} (dR_T/dx)$. The term P_0 representing the static pressure in the magnetosheath is included in the present formulation since $Knmv^2 \sin^2 \alpha$ is expected to be low on the tail surface where $\sin^2 \alpha \ll 1$. Combining Eqs. (59) and (60) and using $\sin \alpha \approx \tan \alpha \approx dR_T/dx$, one obtains

$$\frac{dR_T}{dx} = \frac{1}{M}\left[\left(\frac{R_*}{R_T}\right)^4 - 1\right]^{1/2} \tag{61}$$

which has the quadrature solution

$$\int_{R_0/R_*}^{R_T(x)/R_*} \frac{dr}{(r^{-4} - 1)^{1/2}} = \frac{x - x_0}{MR_*} \tag{62}$$

where $M^2 = Knmv^2/P_0$ and $R_* = (2F_T^2/\mu_0\pi^2 P_0)^{1/4}$ is the asymptotic radius of the tail at the end of the flaring region (where $dR_T/dx \to 0$). Flaring ceases at a finite distance x_* downstream given by

$$\frac{x_* - x_0}{MR_*} = \int_{R_0/R_*}^{1} \frac{dr}{(r^{-4} - 1)^{1/2}} \simeq 0.6 - \frac{1}{3}\left(\frac{R_0}{R_*}\right)^3. \tag{63}$$

For typical solar wind values of $P_0 = 1.7 \times 10^{-11}$ Newton/m^2 and $M \simeq 9$, we have $R_* \simeq 15R_E(F_T)^{1/2}$ where F_T is in units of 10^8 Wb.

In Eq. (62) $R_0 = R_T(x_0)$ is the initial radius of the tail at the downstream position x_0 where the tail solution is first valid. In the region $x \ll x_*$ the integration of Eq. (62) can be simplified since $r^{-4} \gg 1$ and

$$\frac{R_T(x)}{R_0} = \left(1 + \frac{x - x_0}{L}\right)^{1/3} \tag{64}$$

and

$$B_x(x) = \varepsilon \frac{2F_T}{\pi R_0^2}\left(1 + \frac{x - x_0}{L}\right)^{-2/3} \tag{65}$$

where $L = MR_0^3/3R_*^2$ is the flaring tail scale length. Since

$$\frac{\Delta L}{L} = 3\frac{\Delta R_0}{R_0} - \frac{\Delta F_T}{F_T} \tag{66}$$

it follows from Eqs. (64) and (65) that at distances where $x - x_0 \gg L$

$$\frac{\Delta R_T}{R_T} = \frac{1}{3}\frac{\Delta F_T}{F_T} \tag{67}$$

and

$$\frac{\Delta B_x}{B_x} = \frac{1}{3} \frac{\Delta F_{\mathrm{T}}}{F_{\mathrm{T}}} \tag{68}$$

so that the increase in F_{T} is due equally to increases in the diameter of the tail and in the field strength in the tail lobe. Although the assumption of the constancy of the flux content is rather crude, the preceding model thus reproduces the growth-phase behavior observed in the tail lobe at $|x|$ of $15 \sim 80R_{\mathrm{E}}$ rather satisfactorily.

The degree of compression of the plasma sheet would be greater in its near-earth part than in the distant part, since it is in the near-earth region where the enhanced solar-wind pressure is exerted on the flared magnetopause. Accordingly, the enhancement of the cross-tail current as well as the antiearthward displacement of field lines would be greater there. Hence in the earthward regions of the plasma sheet as well as on the earthward side of its inner boundary, the magnetic field is reduced as the plasma sheet becomes compressed; the reduction in the field magnitude would result also from the sunward drift of the field lines induced by the day-side reconnection process (namely, by the effect of the line-tying current). Due to the frozen-in condition and the conservation of the first two adiabatic invariants the weakening of the magnetic field leads to the reduction of the plasma density and temperature. Hence in order to maintain balance with the enhanced magnetic pressure exerted from the tail lobe, the plasma in the near-earth portion of the plasma sheet has to reduce its dimension, namely, the plasma sheet has to thin in that region. This probably explains why the earthward portion of the plasma sheet is severely thinned during the growth phase. It is interesting to note that the radial distance of $\sim 15R_{\mathrm{E}}$ beyond which an appreciable thinning is not usually observed during the growth phase roughly marks the position beyond which β at the midplane of the tail is expected to exceed 1. (The thinning of the plasma sheet during the growth phase is very much exaggerated in Figure 65, partly because the distant neutral line is depicted at an arbitrary position rather close to the earth.)

Relaxation by Near-Tail Reconnection

The change in the tail field configuration that is observed at the onset of the expansion phase is due to severe weakening, or disappearance, of the earthward portion of the cross-tail current. This suggests that the cross-tail current becomes unstable when the thickness of the plasma sheet reaches some critical level. Since the cross-tail current is of the nature of the j_P current that depends on the plasma pressure, the instability concerned should be the one that is associated with a change in the distribution of the pressure. Such instabilities as the two-stream instability and the ion-acoustic instability, which may be of much significance for ionospheric and field-aligned currents that are of the nature of the conduction current $j_C = \sigma E$, are not likely to play a significant role in the formation of the neutral line inside the plasma sheet. Schindler (1974) suggested that the ion-tearing mode would set in when the characteristic time scale for the development of this instability becomes less than the gyroperiod of protons associated with the normal component B_n of the magnetic field at the midplane of the plasma sheet. In that circumstance the gyroscopic motion of protons, which is the basis of the proton contribution to j_P, becomes seriously perturbed. This condition can be expressed as

$$\gamma > \Omega_{i,n}/2\pi \tag{69}$$

where $\Omega_{i,n} = eB_n/m_p$ and γ, the growth rate of the ion-tearing mode instability, is given by

$$\gamma \sim (\pi)^{1/2} \frac{v_{th,i}}{L_z} \left(\frac{a_i}{2L_z} \right)^{3/2}$$

where a_i and L_z represent the proton gyroradius and the half-width of the plasma sheet. $T_i > T_e$ is assumed. The preceding condition is plotted in the $L_z - B_n$ plane in Figure 66. The numerical factor ζ (≥ 1) is introduced to take into account the residual electron damping plus the fact that it may take several linear growth times for an initial perturbation to grow to macroscopic amplitudes. Although the minimum thickness of $\sim 1 R_E$ and the magnetic field of $B_z \sim 10\gamma$, $B \sim 20\gamma$ frequently observed by OGO 5 at $|x| \lesssim 10 R_E$ toward the end of the growth phase are on the stable side of the diagram, the satellite was obviously well earthward of the neutral line because B_z was so high, and at slightly greater distances both B_z and L_z may have become sufficiently small to satisfy the condition of the ion-tearing instability (Nishida and Fujii, 1976).

The speed of the thinning of the plasma sheet is estimated to be about 5 km/s because there is a delay of $1 \sim 2$ h between the sudden southward turning of IMF (which presumably causes a sharp increase in the rate of flux transport to the tail) and the onset of the expansion phase (Russell, 1974). The speed would be augmented if the overall magnetosphere is

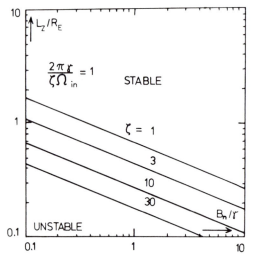

Fig. 66. Plots of the condition for the onset of the ion-tearing mode in the $L_z - B_n$ plane. $v_{th,i} = 500$ km/s and $B = 20\gamma$ are used for computing γ (Schindler, 1974)

suddenly compressed. Such a compression would be particularly effective if it occurs when the plasma-sheet thickness has already reached the marginal state of the instability, and in such circumstances it may swiftly bring the plasma-sheet structure into the unstable range. This is likely to be the reason why SIs tend to trigger the expansion phase when they occur during intervals of southward IMF and/or geomagnetic agitation.

The reconnection process is associated, intrinsically, with the conversion of the magnetic field energy to the particle kinetic energy. When the reconnection is in progress and the magnetic field is changing its strength ($\partial B/\partial t \neq 0$), there is an induced electric field E_{ind} in the reconnection region. The change in particle kinetic energy is given by the sum of the gyro-betatron effect which is proportional to $\partial B/\partial t$ and the drift-betatron effect which is proportional to E_{ind} (Roederer, 1970). The first term acts to increase the kinetic energy in regions where the magnetic field is increasing. The strength of the induced electric field would be the largest at the formation of the near-tail neutral line when the magnetic field changes rapidly. If the magnetic field changes by 10γ in 10 s over the dimension of $1R_E$, the induction electric field of 6 mV/m (2×10^{-7} statvolt/cm) can result. Particles gain energy from this electric field by moving parallel to it in the neighborhood of the neutral line. The mode of the energy gain depends on the degree of the particle scattering in that region. If the plasma is sufficiently turbulent and its pitch-angle distribution is kept nearly isotropic, the acceleration takes the form of heating. If, on the other hand, the scattering mechanism is weak, particles would propagate in the direction of the electric force and attain high energy.

In the former case a significant dawn-dusk asymmetry would not arise in the distribution of heated protons and electrons, while in the latter case of the weak scattering, the energized protons would concentrate toward the dusk-side magnetopause while energized electrons move toward the dawn-side magnetopause, and the electric field is expected to be modified by the resulting polarization effect. As an illustration, Figure 67 shows proton trajectories in a model magnetotail where the uniform magnetic field, whose direction is perpendicular to the plane of the sheet, reverses stepwise across the neutral sheet at $z = 0$ and the electric field is directed toward the left. The tendency of protons to accumulate toward the region of the low electric potential is clearly seen. (In this model calculation the electric field is strongly modulated by the polarization since the electric discharge via the ionosphere is not taken into account.) (Cowley, 1973). Even when the efficiency of the scattering mechanism is appreciable, protons in the high energy tail of the population may not be strongly affected thereby and the suprathermal ion population may be produced with a distinct dawn–dusk asymmetry. This may be considered to be the reason why the dawn–dusk asymmetry is observed in energetic proton bursts but not in energetic electron bursts.

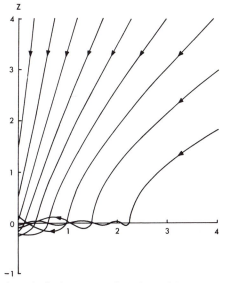

Fig. 67. Proton trajectories in the cross section of a model magnetotail (Cowley, 1973)

The velocity u_R of the field-line flow toward the neutral line, averaged over the duration T of the reconnection, is given roughly by $u_R = \Delta F_T/BDT$, where ΔF_T is the excess flux that is reconnected. Using $\Delta F_T = 2 \times 10^8$ Wb, $T = 30$ min, D(the tail diameter) $= 40R_E$ and $B = 10\gamma$, we get $u_R = 40$ km/s and the associated electric field of 0.4 mV/m. This value of u_R compares favorably with the upper limit $\sim 0.1V_A$ of the reconnection speed predicted by the steady reconnection model of Petschek (1964) if we use $n = 1/\text{cm}^3$ as the number density of plasma in the region of $B = 10\gamma$. The actual instantaneous strength of the electric field could be higher than the foregoing, however, because turbulent electric fields may have high intensities around the reconnection region.

We wish to emphasize at this point that while most of the existing theories for the reconnection and neutral-line processes have been developed for steady-state conditions, the near-tail reconnection characterizing the expansion phase of substorms is basically an unsteady phenomenon. This means, for example, that contrary to ordinary assumptions the flow field can be divergent or convergent and the electric field is not expressible by an electric-potential field. It seems important to keep this difference in mind when comparing substorm observations with available reconnection theories.

Current Circuit of Magnetic Substorm

During the expansion phase the weakening of the near-earth portion of the cross-tail current occurs in a wide longitudinal sector but does not develop over the entire width of the tail; the current flowing in the neighborhood of the tail magnetopause seems to stay essentially intact. The resulting divergence of the current would give rise to an electric polarization at longitudes bounding the region of the reduced current intensity. By virtue of the high electric conductivity that is normally found along field lines, this polarization would be projected to the ionosphere and drive the westward electrojet that acts to close the circuit. Hence it has been envisaged that a three-dimensional current circuit would be produced in the magnetotail–ionosphere system as depicted on the night side of Figure 68a as currents a, b, c, and d (e.g., Kamide et al., 1976). The ground magnetic effect of this circuit has been calculated by using increasingly sophisticated models (e.g., Horning et al., 1974), and basic substorm signatures in the geomagnetic field such as the high-latitude negative bay and the low-latitude positive bay have been attributed to this tail–ionosphere current circuit and the associated ionospheric Hall current.

Another three-dimensional current system depicted inside the tail–ionosphere current system in Figure 68a as currents a', b', c', and d' is considered to be due to the injection of the tail plasma toward the inner magnetosphere. As will be discussed in the next Chapter the magneto-spheric part (labeled partial ring current) of this current system is due to the j_P current flowing in the injected plasma. The current system closes via the ionosphere as polarization charges are produced on the lateral (i.e., eastward and westward) edges of the injected plasma due to the differential motion of protons and electrons. It has been suggested that the partial ring current starts to develop during the growth phase as the inner boundary of the plasma sheet moves gradually inward (being car-ried by the sunward drift induced by the day-side reconnection) and thus causes the H-component decrease observed at low-latitude ground sta-tions during that phase. During the expansion phase the H-component decrease is intensified on the evening side, namely the low-latitude nega-tive bay is intensified, as the plasma is rapidly injected from the tail. This effect is not evident in the midnight to morning sector, however, because the H-component increase produced by the tail-ionosphere circuit has a dominant influence there (Crooker and McPherron, 1972).

The ionospheric current flow corresponding to the above is shown in Figure 68b. According to this picture the westward electrojet is formed by the longitudinal closure of the field-aligned currents flowing in and out along the poleward circuit, while the eastward electrojet is produced as the meridional closure of the current takes a detoured route in the

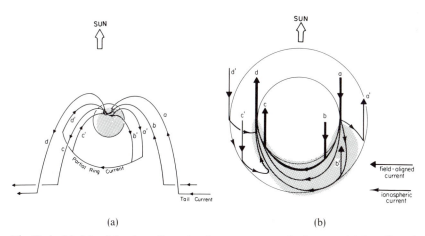

(a)   (b)

Fig. 68a,b. Models of the three-dimensional current system of substorms. (a) Coupling of the magnetospheric and the ionospheric current systems, and (b) distribution of the current in the auroral oval. The dotted area represents the region of the westward electrojet (Kamide et al., 1976)

premidnight region. A current system like this one has been obtained by a model calculation in which the field-aligned current is assigned at boundaries of a circular zone of enhanced conductivity (Yasuhara et al., 1975). However, in order to make the picture more complete it is probably necessary to incorporate the outflow of the current from the Harang discontinuity. Note also that the ionospheric Hall current that closes within the ionosphere is not included in the Figure and has to be added when a comparison with the geomagnetic disturbance field is attempted. The effect of the field-aligned current on the ground magnetic field has been examined extensively by Fukushima (1976).

The appearance of DP2 during the growth phase seems natural as expansion phase is to follow the reconnection on the dayside magnetopause. The distinction between geomagnetic effects of the day-side mag-

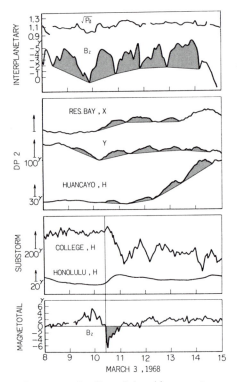

Fig. 69. Comparison of geomagnetic effect of day-side magnetopause and night-side tail reconnections

netopause and the night-side tail reconnections is clearly seen in the example in Figure 69. The top panel shows the dynamic pressure of the solar wind and the solar-magnetospheric z component of the magnetic field. This B_z-record suggests that the day-side nose reconnection rate varied quasi-periodically with a period of about one hour, and the consequence of this variation can be recognized as DP2 in the polar-cap (Resolute Bay at 83°) and the equatorial (Huancayo at $-1°$) magnetograms reproduced in the second panel. The bottom panel, on the other hand, displays a record of the solar-magnetospheric z component of the tail magnetic field observed in low latitudes at $x = -33R_E$, and this record suggests that the reconnection probably operated at the near-earth neutral line once for about 45 min during the 7-h interval examined here. The onset of this reconnection is associated with the observation of the expansion-phase signatures at College (65°) and Honolulu (21°) around midnight. The disturbance field of the DP2 mode apparently continued through the expansion and the recovery phases of the substorm.

The auroral-zone magnetic disturbances of the recovery phase persist after the z component of the low-latitude tail field in the near-earth region has returned to the northward range (e.g., Fig. 69). These disturbances would comprise such elements as the DP2 field and the disturbance associated with the neutral line that has shifted to the distant tail. It is also conceivable that the energy that has been introduced to the outer radiation belt (including the ring current region) during the expansion phase keeps driving electric currents in the auroral-zone ionosphere.

III.5 Magnetosphere–Ionosphere Coupling During Substorms

As discussed in the last Section, three-dimensional current systems are generated during substorms as the cross-tail current is channeled to the ionosphere. This is associated with the formation of the electric field in the magnetosphere–ionosphere system. Let us look more closely at the nature of such magnetosphere–ionosphere coupling.

Basic Principles

From the electrodynamic point of view the distinction between the ionosphere and the magnetosphere arises from the difference in the ratio of the ion gyrofrequency to the collision frequency. The equations of motion of electrons and ions are essentially given by

$$\frac{dv_e}{dt} = -\frac{e}{m_e} E - \frac{e}{m_e}[v_e \times B] - \frac{1}{nm_e}\nabla p_e - v_e[v_e - u] \tag{70}$$

$$\frac{dv_i}{dt} = \frac{e}{m_i} E + \frac{e}{m_i}[v_i \times B] - \frac{1}{nm_i}\nabla p_i - v_i[v_i - u] \tag{71}$$

when pressure is isotropic, gravity is negligible, and ions are singly ionized. v and u are collision frequency with neutral constituents and mean velocity of neutrals, respectively, and all other variables have the usual meanings. When the effect arising from the neutral wind (namely, the ionospheric wind dynamo) is set aside, u can be equated to zero. In the magnetosphere above about 160 km, the collision frequency (including, strictly speaking, the effect of ion-electron collisions) is low compared with the gyrofrequency, namely,

$$v_e < \Omega_e \left(=\frac{eB}{m_e}\right), \qquad v_i < \Omega_i \left(=\frac{eB}{m_i}\right),$$

and from Eqs. (70) and (71) it follows that

$$j_\perp = ne(v_i - v_e)_\perp$$

$$\doteq \frac{1}{B^2}[B \times \nabla p] + \frac{nm_i}{B^2}\left[B \times \frac{dv_i}{dt}\right] \tag{72}$$

where the first term is equal to j_P of Eq. (11) while the second term becomes j_A of Eq. (57) if $|\partial v_i/\partial t| \ll |(v_i \cdot \nabla)v_i|$.

In the ionosphere, on the other hand, where the collisions cannot be neglected, Eqs. (70) and (71) lead to

$$j = \sigma_0 E_\parallel + \sigma_P E_\perp + \sigma_H \frac{E_\perp \times B}{B} \tag{73}$$

where σ's represent parallel, Pedersen and Hall conductivities:

$$\sigma_0 = \frac{n_e e}{B}\left(\frac{\Omega_i}{v_i} + \frac{\Omega_e}{v_e}\right) \tag{74a}$$

$$\sigma_P = \frac{n_e e}{B}\left(\frac{\Omega_i v_i}{\Omega_i^2 + v_i^2} + \frac{\Omega_e v_e}{\Omega_e^2 + v_e^2}\right) \tag{74b}$$

$$\sigma_H = \frac{n_e e}{B} \left(\frac{\Omega_i^2}{\Omega_i^2 + v_i^2} - \frac{\Omega_e^2}{\Omega_e^2 + v_e^2} \right). \tag{74c}$$

(σ_H given above takes on negative values in the ionosphere so that $|\sigma_H|$ is usually tabulated as the Hall conductivity.) It is customary to assume that $E_\perp$ does not change much with height within the ionosphere and to use Eq. (73) in the integrated form:

$$J_\perp = \Sigma_P E_\perp + \frac{\Sigma_H}{B} E_\perp \times B \tag{75}$$

where Σ_P and Σ_H are integrals of σ_P and σ_H over the ionospheric layer. Examples of daily variations of Σ_P and Σ_H in the auroral zone, derived from the electron density profile obtained by an incoherent scatter experiment and an appropriately chosen neutral atmosphere profile, will be presented later.

The field-aligned current arises from the divergence of these currents in the plane perpendicular to the magnetic field. Let us denote $j_\perp$ given by Eqs. (72) and (73) by j_M and j_I. The magnetospheric contribution to the field-aligned current is given by

$$\boldsymbol{V}_\perp \cdot \boldsymbol{j}_M = (\boldsymbol{j}_{D1} + \boldsymbol{j}_{D2}) \cdot \boldsymbol{V}_\perp p/p + \boldsymbol{V}_\perp \cdot \left(\frac{\rho}{B^2} \left[\boldsymbol{B} \times \frac{d\boldsymbol{v}_i}{dt} \right] \right) \tag{76}$$

where

$$\boldsymbol{j}_{D1} = \frac{p(\boldsymbol{B} \times \boldsymbol{V}B)}{B^3} \qquad \text{(gradient-}B\text{ current)}$$

$$\boldsymbol{j}_{D2} = \frac{p[\boldsymbol{B} \times (\boldsymbol{b} \cdot \boldsymbol{V})\boldsymbol{b}]}{B^2} \qquad \text{(curvature current)}$$

and

$$\boldsymbol{b} = \boldsymbol{B}/B, \qquad \rho = nm_i.$$

For example, when the plasma-sheet thinning occurs in a limited longitudinal sector, $\boldsymbol{V}_\perp p$ is produced, and its product with the gradient-B and the curvature currents flowing across the tail acts as a source of the field-aligned current through the first term of Eq. (76). When the perturbation in $\boldsymbol{B}$ is small compared with $|\boldsymbol{B}|$, while the perturbation in $\boldsymbol{E}$ is comparable to $|\boldsymbol{E}|$ itself, the second term of Eq. (72) can be simplified as (Atkinson, 1970)

$$\frac{\rho}{B^2}\left[B \times \frac{dv_i}{dt}\right] \doteq \frac{\rho}{B^2}\left[B \times \frac{d}{dt}\left(\frac{E \times B}{B^2}\right)\right]$$

$$\doteq \frac{\rho}{B^2}\frac{dE_\perp}{dt}$$

$$\doteq \frac{\rho}{B^2}\frac{\partial E_\perp}{\partial t} + \sigma_M E_\perp \qquad (77)$$

where

$$\sigma_M = \frac{\rho|E_\perp|}{B^3 d}$$

d being the characteristic length of the electric field. Since slow variations are usually considered, the $\partial E_\perp/\partial t$ term will be neglected. On the other hand, ionospheric contribution to the field-aligned current results from

$$V_\perp \cdot j_I = j_I \cdot V_\perp n/n + \sigma_P V_\perp \cdot E_I \qquad (78)$$

where E_I is the ionospheric electric field. [$\nabla \cdot (E \times B)$ is zero when both E and B represent potential fields.] Evidently a horizontal nonuniformity of the ionospheric electron density gives rise to the field-aligned current through the first term of Eq. (78).

The overall equation of continuity of the three-dimensional circuit is obtained by integrating Eqs. (76) [with (77)] and (78) over relevant ranges of altitudes along field lines and equating their sum to zero (Sato, 1976);

$$(\Sigma_P + \Sigma_M) V_\perp \cdot E_I = Q_I + Q_M \qquad (79)$$

where Σ_M is the field-aligned integrated value of σ_M and

$$Q_M = -s(J_{D1} + J_{D2}) \cdot V_\perp p/p$$
$$Q_I = -J_I \cdot V_\perp n/n,$$

s being a scale factor that represents the ratio of the cross-sectional area of a tube of force in the magnetosphere to that in the ionosphere. Capital letters represent the integrated values of the parameters expressed by corresponding small letters. If $\Sigma_P \gg \Sigma_M$, Eq. (79) can be interpreted to mean that the principal function of the field-aligned current is to discharge the polarization produced in the magnetosphere via the ionospheric path. The process is associated with the formation of the electric field. The polarization charges produced at ionospheric conductivity discontinuities, on the other hand, cannot be neutralized by the magnetospheric circuit. Swift depletions of the ionospheric polarization require $\Sigma_P \ll \Sigma_M$. Physically, $\Sigma_M E$ expresses the current that flows to accelerate the magnetospheric medium by the Lorentz force to the assigned drift velocity of $(1/B^2)[E \times B]$, and Σ_M is larger the more massive the medium is. If we take $n = 1/\text{cm}^3$, $E_M = 1\,\text{mV/m}$ (3×10^{-8} statvolt/cm), $B = 10\gamma$ and take d equal to the width of the active region in the magnetosphere, we get $\Sigma_M \sim 1$ mho (10^{12} statmho). In comparison, the integrated Pedersen conductivity in the auroral zone is several mhos in active nighttime conditions (Brekke et al., 1974).

Observations of Field-Aligned Currents

An example of the observation of the field-aligned current at the bottom of the magnetosphere is shown in Figure 70. This example was obtained in the premidnight region by the Triad satellite, which was cruising at an altitude of about 800 km. The density of the field-aligned current is inferred from its effect on the local magnetic field as illustrated in Figure 70a; the record of magnetic perturbations is shown at the bottom and the inferred density of the field-aligned current is shown in the middle. Usually there is little perturbation parallel to the main field, and a much greater perturbation is observed in the east–west component than in the north–south component, indicating that the field-aligned currents flow in east–west-aligned sheets. The observation was made during a substorm expansion phase whose onset was recorded at 0950 UT. The dominant features of the field-aligned current are: an outward flowing current covering geomagnetic latitudes of $63.8° \lesssim \Lambda \lesssim 65.2°$ and an inward current flowing on its equatorward side. The outward current was about twice as strong as the inward current, and the poleward border of the field-aligned current was marked by the discrete arcs at the poleward border of the region of the auroral luminosity. Figure 70b presents the latitude profile of ground magnetic perturbation recorded in the neighborhood of the projection of the satellite trajectory a few minutes after the passage of the satellite, and it is clearly seen that the intense outward current was flowing out from the equatorward portion of the westward electrojet (Armstrong et al., 1975). Although not always clear, there tends to be an eastward electrojet in the region of the inward field-aligned current of the premidnight sector (Rostoker et al., 1975; Kamide et al., 1976).

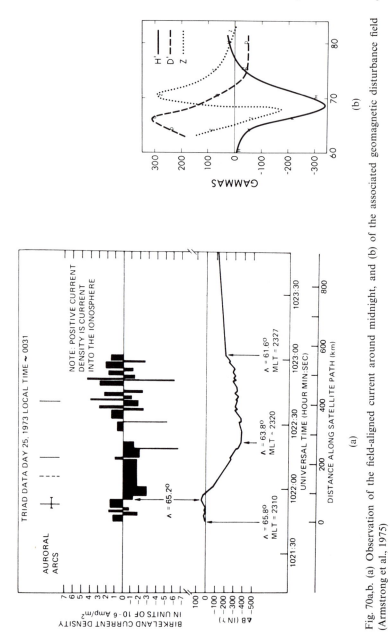

Fig. 70a,b. (a) Observation of the field-aligned current around midnight, and (b) of the associated geomagnetic disturbance field (Armstrong et al., 1975)

Since the satellite traversed the auroral oval during an expansion phase before midnight when the preceding example was obtained, it is tempting to identify the dominant outward current with the premidnight portion of the tail-ionosphere current circuit illustrated in Figure 68a. Although this interpretation may eventually turn out to be correct, at the time of this writing, the analysis of the field-aligned current has not yet been advanced to allow us to interpret the observations uniquely in terms of substorm phases. Moreover, the observations of field-aligned currents (frequently referred to as Birkeland currents) have revealed complexities that are not incorporated in the simplified picture of Figure 68a. Hence we shall describe the overall characteristics of the field-aligned current that have been derived by statistical means. Figure 71a is a summary of the distribution of field-aligned currents during quiet intervals ($|AL| < 100\gamma$). Directions of the currents are basically antisymmetric with respect to the noon–midnight meridian. Except during two intervals of a few hours' span, of which one is around midday and another before midnight, the distribution consists of two circumpolar belts that follow the auroral oval. The current density is about two times stronger in the poleward belt than in the equatorward belt in both morning and evening sectors. (The total current flowing into or out of the poleward belt is several times of 10^6 A.) As the magnetic activity increases, the current density increases in both latitude belts and in both local-time sectors while the entire belt moves equatorward (Iijima and Potemra, 1976a and b).

In the poleward belt of the field-aligned current, the current flows into the ionosphere in the morning sector and flows out in the evening sector. In the equatorward belt, on the other hand, the current flows into the ionosphere in the evening sector and flows out in the morning sector. The continuation of the field-aligned currents to the night-side magneto-sphere has been detected at distances of several R_E and more, and according to these observations the poleward current belt continues to the high-latitude boundaries of the plasma sheet while the equatorward current belt is connected to the region where the magnetic field is still nearly dipolar. The field-aligned currents on the plasma-sheet boundaries have been found to exist at nearly all times but are intensified during substorms (Sugiura, 1975).

Since the field-aligned current has been found to exist even during quiet intervals, the disruption of the cross-tail current in the near-tail region as depicted in Figure 68a cannot be considered as the sole source of the field-aligned current in the poleward belt. It would be that the currents associated with the day-side reconnection (and hence DP2) and with the slow reconnection operating at the neutral line in the distant

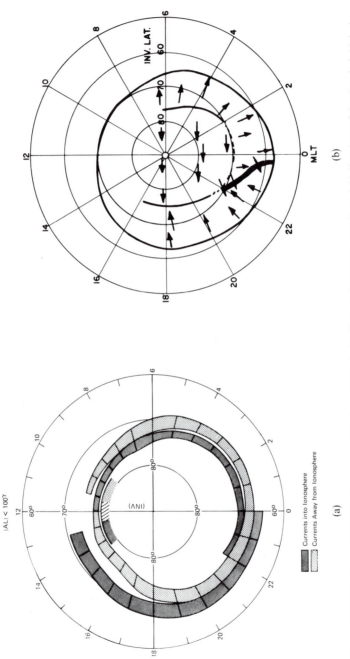

Fig. 71a,b. (a) Distribution of the field-aligned current during quiet periods (Iijima and Potemra, 1976b), and (b) directions of the electric field during moderately disturbed to disturbed intervals (Maynard, 1974)

tail represent the nature of the field-aligned current in quiet times. In addition, Figure 71a shows a very high latitude region of the field-aligned current having a few-hours' span around midday. Directions of the current flow in this region are opposite to those observed in the 'poleward belt' located immediately equatorward, and they are consistent with the disturbance field of Figure 43b, which is correlated with the northward IMF. A point of an apparent disagreement, however, is that the field-aligned current in this very high latitude belt has been reported to intensify as the southward IMF component increases (Iijima and Potemra, 1976b). The nature of this third region of the field-aligned current also remains to be clarified.

Electric Field and Conductivity in the Ionosphere

The overall distribution of the high-latitude electric field is represented schematically in Figure 71b (Maynard, 1974). The principal features are; the dawn-to-dusk field in the polar cap, the equatorward field in the region of the westward electrojet covering the premidnight to morning sector, and the poleward field in the region of the eastward electrojet covering the evening to premidnight sector. In the premidnight region of a few hours' duration the latter two regions latitudinally overlay. Thus the Harang discontinuity noted in the equivalent current system appears to have a corresponding feature in the electric field (Wescott et al., 1969; Maynard, 1974). The overlay structures of the field-aligned current (Figure 71a) and of the electric field (Figure 71b) noted at the Harang discontinuity probably represent closely linked features; the convergence of the electric field at the Harang discontinuity (due to the presence of the equatorward field on the poleward side and of the poleward field on the equatorward side) means the convergence of the horizontal Pedersen current, and the field-aligned current is expected to flow out of the discontinuity unless the Hall conductivity changes across the discontinuity in such a way that the divergence of the Hall current makes up for the convergence of the Pedersen current. The electric field distribution of Figure 71b exists both during substorms and in quiet times and the entire structure moves to lower latitudes as activity increases.

The enhancement in the ionospheric conductivity at the auroral oval should certainly be among the basic elements that influence the character of the auroral electrojet, because the intensification of the precipitating particles is one of the characteristic features of the substorm phenomenon. Figure 72 shows simultaneous observations of height-integrated conductivities, electrostatic electric fields, and dynamo electric fields obtained

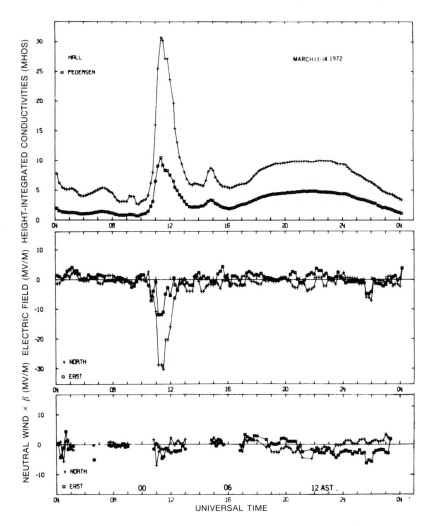

Fig. 72. Conductivity and electric field obtained by the Chatanika incoherent scatter radar
(Brekke et al., 1974)

by the incoherent scatter method at Chatanika (65°). Here the 'electrostatic field' is the electric field observed in the observer's frame of reference; although $\partial B/\partial t$ is obviously not zero during disturbances, rot $E = 0$
is a good approximation since $(\Delta B)L/(\Delta E)T \sim 10^{-1}$ for typical values of
$\Delta B \sim 10^3\gamma$, $L \sim 300$ km, $\Delta E \sim 10$ mV/m (3×10^{-7} statvolt/m), and
$T \sim 5$ min. The 'dynamo electric field' is the field originating from the
neutral wind motion with velocity u across the magnetic field and is

given by $u \times B$. Since the conductivity tensor has been derived in the frame of reference fixed to the neutral gas medium [namely, u was set equal to zero when Eqs. (70) and (71) were solved to yield (74)], this electric field has to be included for deriving the ionospheric electric current (see Section IV.5).

When the attention is focussed on the midnight part of the diagram, a pronounced peak is seen in both Hall and Pedersen conductivities starting from ~ 1030. That Σ_H is more enhanced than Σ_P can be interpreted to mean that the conductivity enhancement is produced mainly by electrons with energies greater than 5 keV that penetrate below the 125 km altitude. The electrostatic field is also intensified and directed southwest during the corresponding interval, whereas the dynamo electric field is much weaker than the electrostatic field. The negative bay that is observed in the H component of the magnetic field at College, which is located near Chatanika, closely follows the east–west component of the ionospheric current derived from observed values of conductivity and electric field (Brekke et al., 1974). The enhancements of the southwest electric field during the substorm expansion phase have been observed repeatedly also by the antennae lifted by balloons at geomagnetic latitudes of 66° to 70° Mozer, 1971).

An increase in the northward component of the electric field in the region of the eastward electrojet is illustrated by Figure 73a. There occurred on this day an expansion phase from 0924 according to the Honolulu magnetogram. In this substorm the development of the westward electrojet was initiated far poleward of the Chatanika area as seen by magnetograms from the Alaskan chain (Fig. 73b). At Poker Flat near Chatanika the H-component perturbation was positive until ~ 0940, and correspondingly the electric field at Chatanika had a positive northward component. The reversal of this component occurred around 1000 when the westward electrojet fully extended to the Chatanika area as demonstrated by the southward turning of the H-component record at Poker Flat. This equatorward shift of the westward electrojet can be interpreted to represent the equatorward shift of the Harang discontinuity due either to the local-time dependence of its latitude or to the equatorward shift of the entire structure with an increasing activity.

Observations like Figure 72 may be considered to suggest a possibility that the conductivity enhancement is the principal cause of the expansion-phase onset. The following scheme, for example, has been suggested by Coroniti and Kennel (1972a). The enhanced ionospheric conductivity intensifies the ionospheric current and, by virtue of the divergence or convergence of the current at boundaries of the region of

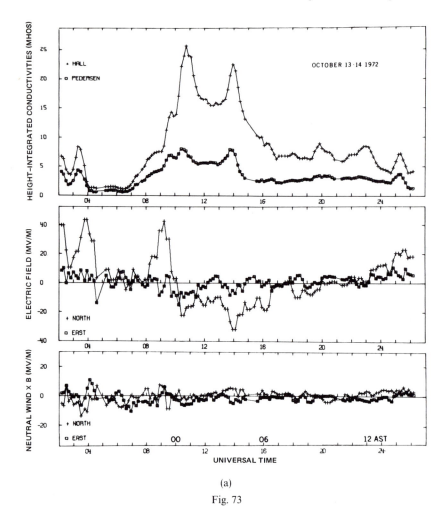

(a)

Fig. 73

the enhanced conductivity, the field-aligned currents are also intensified. If the intensification becomes strong enough to create a state of plasma turbulence out of instability, the current along field lines would become blocked by the resulting increase in the anomalous resistivity. Then divergence/convergence of the ionospheric current results in the polarization electric field, which acts to intensify the ionospheric hall current. When the geometry of the system is assumed appropriately, it is possible to compare this intensification with the appearance of the intense westward electrojet. However, the observations of field-aligned currents do not seem to demonstrate the predicted blocking effect during expansion phases. Instead, the density of the field-aligned current is reported to

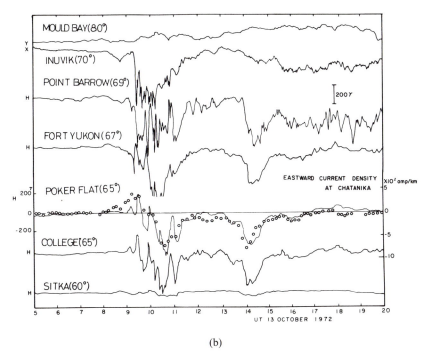

(b)

Fig. 73a,b. (a) Chatanika incoherent scatter data, and (b) the corresponding ground magnetograms. In (b) the eastward current density at Chatanika estimated from the incoherent scatter data (circles) is compared with the H-component record at Poker Flat (Brekke et al., 1974; Kamide and Brekke, 1975)

increase with Kp (Iijima and Potemra, 1976a). Hence it is doubtful that the conductivity enhancement is indeed the triggering factor of the expansion-phase onset.

Nevertheless, it is likely that the polarization electric field is indeed strengthened and helps to intensify the electrojet. To illustrate this, let us take a simple model in which the Hall conductivity is enhanced within an infinitely long strip. The model is intended to approximate the conductivity enhancement at the nightside northern auroral oval, and x, y, z axes point, respectively, southward across the strip, eastward along the strip, and upward perpendicular to the current layer. The magnetic field is assumed to point vertically downward. The height-integrated currents within the strip obey

$$I_x = \Sigma_P E_x + \Sigma_H E_y \tag{80}$$

$$I_y = \Sigma_P E_y - \Sigma_H E_x. \tag{81}$$

[Σ_H used here is $-\Sigma_H$ of the one defined in Eq. (75). In the present definition Σ_H is positive.] It is assumed that the Pedersen and Hall conductivities within the strip are greater than corresponding conductivities outside the strip by Δ_P and Δ_H. Then I_x has nonzero divergence in the plane of the current sheet, and the field-aligned currents I_S and I_N have to flow in z-direction at southern and northern edges of the auroral oval;

$$I_S = \Delta_P E_x + \Delta_H E_y \tag{82}$$

$$I_N = -I_S. \tag{83}$$

The circuit is closed by the magnetospheric current of intensity I_S flowing in the $(-x)$ direction. When Σ_M is used to designate the effective conductivity in the magnetosphere, the magnetospheric circuit can be expressed as

$$I_S = -\Sigma_M E_x. \tag{84}$$

Consequently

$$E_x = -\frac{\Delta_H}{\Delta_P + \Sigma_M} E_y. \tag{85}$$

When the electric field applied from outside the system has only the y component, E_x can be attributed fully to the polarization effect. The current flow along the strip is

$$I_y = \left(\Sigma_P + \frac{\Delta_H \Sigma_H}{\Delta_P + \Sigma_M}\right) E_y. \tag{86}$$

Thus the current along the strip is intensified by the polarization. If conductivities within the auroral oval far exceed those that are available outside, namely, if $\Delta_H \sim \Sigma_H$, $\Delta_P \sim \Sigma_P$ and $\Sigma_M \ll \Sigma_P$, it follows that

$$I_y = \left(\Sigma_P + \frac{\Sigma_H^2}{\Sigma_P}\right) E_y. \tag{87}$$

Since $\Sigma_H/\Sigma_P > 1$, the second term in the parenthesis is larger than the first and its importance becomes more pronounced when the Σ_H/Σ_P ratio is increased during substorms. The preceding combination of the conductivity is called Cowling conductivity. It has long been envisaged that the Cowling current plays an important role in the formation of the electrojet (Boström, 1964; Fukushima, 1969). From Eqs. (84) and (85) the field-aligned current is given by

$$I_S = \frac{\Delta_H \Sigma_M}{\Delta_P + \Sigma_M} E_y. \tag{88}$$

Thus under the westward electric field ($E_y < 0$) as observed during the 1030 ~ 1230 interval in Figure 72, an intense westward electrojet is expected to flow inside the high conductivity strip by virtue of the formation of the strong polarization field that is directed southward, and produce a large decrease in the H component (i.e., a negative bay) underneath. The association of the westward electrojet with the southwestward electric field inside the region of enhanced conductivity is indeed consistent with the observation. On the other hand, the associated flow of current along the field line is expected to be downward ($I_S < 0$) at the equatorward edge of the high conductivity strip due to Eq. (88). At this point the agreement between the model and the observation breaks down; as seen in Figure 70, the field-aligned current that flows at the equatorward portion of the westward electrojet is directed upward. What this disagreement tells us is that we cannot attribute the origin of the field-aligned current at the equatorward border of the westward electrojet to the combined action of the westward electric field and the conductivity enhancement in the ionosphere. Although it is likely that the Cowling conductivity contributes to the formation of the intense electrojet, the origin of the field-aligned current has to be sought in other mechanisms that operate in the magnetosphere.

Indications of Field-Aligned Electric Field

The field-aligned current that couples the ionosphere with the magnetosphere may be carried either by cold electrons of ionospheric origin or by more energetic particles precipitating from the magnetosphere. Observations of energetic particles have indicated that the electrons precipitating to the dusk side of the auroral oval can be an important carrier of the current flowing upward in that region. Figure 74 shows the integral flux of energetic electrons (0.2 to 25 keV) and the horizontal velocity of the ion drift motion observed by the AE-C satellite at the altitude of

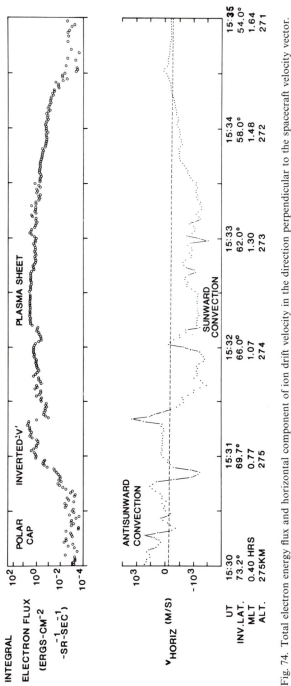

Fig. 74. Total electron energy flux and horizontal component of ion drift velocity in the direction perpendicular to the spacecraft velocity vector. *Dashed base line, lower panel:* $V_{horiz} = 0$ in the corotating frame (Burch et al., 1976)

280 km. The pitch angle distribution (not reproduced) shows that these electrons were precipitating to the ionosphere. The data reproduced were obtained during the traversal of the satellite across the near-midnight part of the auroral oval, and the geomagnetic condition was moderately disturbed (Kp $\sim$ 4). It is seen that the flux of precipitating electrons has a peak (labeled inverted V) in the region that is sandwiched between sharp peaks of sunward and antisunward drifts. Since the geometry of the observation was such that sunward and antisunward drifts corresponded to equatorward and poleward electric fields, respectively, a minimum in div E existed in the region of the velocity shear, and consequently the ionospheric Pedersen current is expected to converge to that region. The observed peak $\sim 10^3$ m/s of the drift velocity corresponds to the electric field of ~ 50 mV/m, and the Pedersen current associated with it is about 0.1 A/m if $\Sigma_P \sim 2$ mho. The current carried upward by the precipitating electrons is, on the other hand, eFl/ε where e is the electronic charge, F is the integral energy flux, ε is the average electronic energy and l is the thickness of the field-aligned current. Using $F \sim 2\pi \times 10^{-3}$ Joule/m^2 s and $l \sim 200$ km that can be read from Figure 74 and also $\varepsilon \sim 3$ keV $\sim 5 \times 10^{-16}$ Joule we obtain $eFl/\varepsilon \sim 0.3$ A/m. Thus the precipitating electrons are quite capable of disposing of the current that converges to the region of the electric field reversal (Burch et al., 1976).

The foregoing structure of precipitating electrons is called 'inverted V' since it shows an inverted V shape in the energy-vs-time spectrogram; as a satellite passes through the structure the electron energy increases, reaches a maximum, and subsequently decreases. This structure has been observed frequently on the premidnight side of the auroral oval in association with the electric-field reversal that occurs at the boundary of the polar cap (Gurnett and Frank, 1973). The energy spectrum obtained in the inverted V clearly shows the deviation from the Maxwellian distribution and suggests that the energy of 1 to ~ 5 keV has been added to particles constituting the higher energy portion of the spectrum. An example reproduced in Figure 75 was obtained over a bright aurora. The energy spectrum has a distinct secondary peak at the energy of ~ 5 keV (Fig. 75b), and the pitch angle of the electrons constituting this peak is highly collimated in the direction of the local magnetic field (Fig. 75a). It is unlikely that a highly collimated, almost monoenergetic flux of electrons such as this can be generated by mechanisms operating in the distant magnetosphere, and the acceleration process seems to be operating in the neighborhood of the ionospheric region where the particle observation was made (Arnoldy et al., 1974). The electric field parallel to the magnetic field is the immediate candidate for the acceleration process,

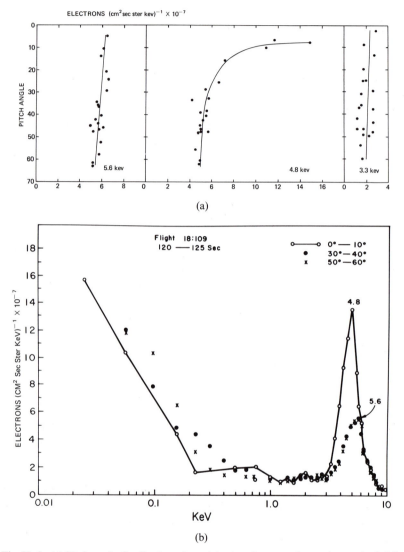

Fig. 75a,b. (a) Pitch angle distribution of precipitating electrons observed over the bright aurora; (b) pitch-angle sorted energy spectra for the same case (Arnoldy et al., 1974)

and supporting this view, the ion observations show that in the inverted V region the precipitating ions are significantly less energetic than in the lower latitude region, and the ratio of integral electron energy flux to integral ion energy flux is higher by $1 \sim 2$ orders of magnitude than elsewhere (Burch et al., 1976).

Hence the electric potential contours in the inverted V region have been suggested to be like Figure 76. At sufficiently high altitudes in the magnetosphere, the field lines are equipotentials with minima at the centers of the contours (left, above), but at low altitude the potential changes along field lines and $E_\parallel$ is present (left, below). Hence a characteristic spatial structure is produced in the energy of electrons (right, above) that precipitate along field lines (Gurnett, 1972). A substantial fraction of the field-aligned potential drop is thought to occur below 2400 km.

Since in the magnetosphere and the upper ionosphere the interparticle collisions are too rare to support the field-aligned electric field by collisional friction, other mechanisms have to be sought to explain the presence of $E_\parallel$, and the following processes are currently being considered as possible candidates (Block and Fälthammar, 1976):

1. In strong magnetic mirrors a hot collisionless plasma can support large magnetic-field aligned electric voltage drops, associated with a differential pitch-angle anisotropy between electrons and positive ions.
2. In the transition region between a hot magnetospheric plasma and a cool ionospheric plasma there may arise large potentials of a thermoelectric nature, even without net current.
3. If imposed electric currents have a direction and magnitude such as to drain the hot magnetosheath plasma of a substantial part of its random electron current, large voltage drops can occur.
4. Instabilities may lead to the generation of waves that inhibit the relative motion of ions and electrons in such a way as to cause anomalous resistivity.

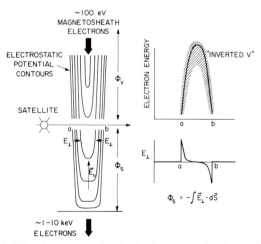

Fig. 76. A model of the electric potential distribution responsible for the inverted V events (Gurnett, 1972)

5. Instabilities may lead to the formation of electric double layers where magnetic field-aligned potential drops occur locally over a small distance.

Both processes 1 and 5 are based on the fact that unless electrons and positive ions have entirely identical pitch-angle distributions at their sources, their number densities become different as they proceed along field lines. Possible source regions are the plasma sheet and the ionosphere. The process was originally examined for the case where the source is only in high altitudes and all particles are mirroring (process 1), but later it was extended to include both ionospheric and plasma-sheet sources and possible streaming of particles along field lines (process 5). In both cases pitch-angle scattering is assumed to be absent and the conservation of the first adiabatic invariant is used as the basis of formulation. The process 4, on the other hand, depends intrinsically on the particle scattering effect of plasma waves. Fälthammar (1977) has estimated the maximum electric field attainable by this process to be about 10 mV/m. Process 2 has been suggested as an extension of the thermoelectric effect to a collisionless medium, but it remains to be shown that particle scattering by plasma waves can play essentially the same role as Coulomb collisions with regard to this effect.

It would be worthwhile to note that these mechanisms that attempt to explain the appearance of $E_{||}$ do not really produce electromotive force. Electric potential drops are originally generated by large-scale processes that involve the entire magnetosphere–ionosphere system, and the $E_{||}$-producing mechanisms work only to apply along field lines some fraction of the entire potential drop. In order that $E_{||}$ be maintained by any of these mechanisms, electromotive force must be generated in the entire system by some other process.

IV. Dynamic Structure of the Inner Magnetosphere

IV.1 Introduction: The Ring Current

During intervals of pronounced geomagnetic activity, called magnetic storm, the low-latitude magnetic field becomes depressed at all local times, indicating the formation of a current circuit that encircles the earth. Since the depression is too great to be attributed to an expansion of the magnetosphere, the ring current is thought to develop within the magnetosphere, and the search for the carrier of such a current led to the prediction of the presence of particles trapped in the geomagnetic field (Singer, 1957).

Figure 77 presents a clear example of the DR field, namely the disturbance field attributable to the ring current. The second through the seventh panels of this Figure reproduce magnetograms from the six low-latitude stations that are nearly equally spaced longitudinally. The solar quiet daily variation, Sq, has been subtracted from each record. It is seen that for more than two days starting from ~ 1000 of 16 February the H components at all the stations are depressed below the quiet-day level. The comparison of these ground observations with the satellite observation made at the radial distance of $6.6R_E$ (reproduced in the top panel) shows an interesting contrast: the field depression starts earlier and ends much faster at $6.6R_E$ than on the ground. This suggests that the ring current was formed initially outside the $6.6R_E$ distance but moved earthward and was located inside $6.6R_E$ for most of its lifetime.

In the preceding example the intensification of DR on the ground is preceded by a distinct SI event about 10 h before. This SI is a clear case of the sudden commencement of a magnetic storm (SSC). The interval between SSC and the development of the world-wide field depression is called 'initial phase' of a magnetic storm, while the interval of the decreasing field is called 'main phase.' 'Recovery phase' is the interval of the gradual recovery of the depressed field.

The nature of the ring current is the j_P current of Eq. (11), and the estimate of the resulting magnetic perturbation necessitates an appropriate knowledge of the spatial distribution and the energy spectrum of

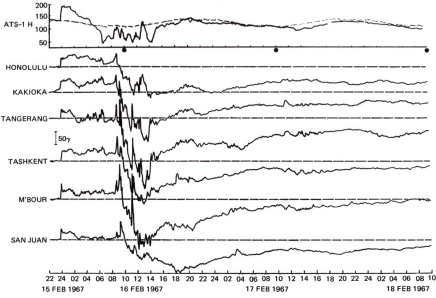

Fig. 77. Comparison of ATS 1 (at r = 6.6R_E) and low-latitude H-component magnetograms during a magnetic storm. The dashed curve superposed on the ATS data is a typical quiet day record. Geomagnetic coordinates of the stations are, Honolulu: 21°, 266°; Kakioka: 26°, 206°; Tangerang: −18°, 175°; Tashkent: 32°, 144°; M'Bour: 21°, 55°; and San Juan: 30°, 3°. Dark dots in the top panel indicate local midnight at ATS-1 (Kawasaki and Akasofu, 1971a)

the plasma injected into the geomagnetic field. Nevertheless, the local-time average I_{DR} of the perturbation field produced on the ground can be expressed by a simple formula:

$$I_{DR} = -\frac{2}{3}\frac{U_T}{U_D}B_D \qquad (89)$$

where U_T is the total kinetic energy of the resident particles, U_D is the energy ($\sim 9 \times 10^{17}$ Joules) of the dipole magnetic field above the earth's surface, and B_D is the dipole field strength on the ground at the equator. This formula was originally derived for particle populations having specific pitch angle distributions but subsequently found to be applicable to arbitrary pitch angle distributions provided that the particle motion is steady (Sckopke, 1966). The formula in fact has been found to be a manifestation of the virial theorem for a closed system in a steady state:

$$2U_T + U_{EM} + U_G = 0 \qquad (90)$$

where U_T is the total kinetic energy, U_{EM} is the total electromagnetic energy, and U_G is the total gravitational energy, and the formula incorporating the magnetopause current contribution has also been obtained from the virial theorem (Siscoe, 1970).

This Chapter deals with the dynamic structure of the inner part of the magnetosphere, starting from the formation of the ring current due to the particle injection from the plasma sheet. The driving agent of the particle motion is the electric field produced by the magnetotail dynamics and by the solar wind–magnetosphere coupling. The effect of the electric field is seen also in the distribution of the cold plasma of ionospheric origin as will be discussed next. The Chapter is completed by a brief description of the geomagnetic quiet daily variations arising from the atmospheric dynamics.

IV.2 Particle Injection from the Plasma Sheet

Energetic particles that inflate the inner magnetosphere originate from the plasma sheet. They are driven from the magnetotail toward the earth by virtue of the drift motion under the intensified electric field.

Injection as a Part of the Substorm Process

The motion of charged particles averaged over gyrations and bounce motions can be expressed by the guiding center velocity:

$$\boldsymbol{u}_\perp = \frac{\boldsymbol{b}}{B} \times \left\{ -\boldsymbol{E} + \frac{\mu}{q} \nabla B + \frac{m}{q} \left(v_\parallel \frac{d\boldsymbol{b}}{dt} + \frac{d\boldsymbol{u}_E}{dt} \right) \right\} \qquad (91)$$

where q and m are charge and mass of the particle, $\boldsymbol{b}$ is the unit vector in the direction of $\boldsymbol{B}$, and $\boldsymbol{u}_E = \boldsymbol{E} \times \boldsymbol{b}/B$ is the drift velocity due to the electric field alone; also, $\mu = \frac{1}{2}mv_\perp^2/B$, where $v_\perp$ is the perpendicular velocity in the frame of reference moving with the guiding center velocity $\boldsymbol{u}_\perp$ (Northrop, 1963). As long as $|\boldsymbol{u}_E|$ does not approach the particle velocity $(v_\perp^2 + v_\parallel^2)^{1/2}$ in that reference frame, $\boldsymbol{u}_\perp$ can be approximated by

$$\boldsymbol{u}_\perp = \frac{\boldsymbol{b}}{B} \times \left\{ -\boldsymbol{E} + \frac{\mu}{q} \nabla B + \frac{m}{q} v_\parallel^2 (\boldsymbol{b} \cdot \nabla) \boldsymbol{b} + \frac{m}{q} v_\parallel \frac{\partial \boldsymbol{b}}{\partial t} \right\}.$$

The second and the third terms in the parenthesis represent so-called gradient-B drift and curvature drift, respectively. Usually the pitch angle distribution of the magnetospheric particles is pancake-shaped; namely, the ratio of particles having larger pitch angles to those having smaller pitch angles is higher than the ratio that would pertain in the case of the isotropic distribution. The anisotropy is produced partly by the loss (by precipitation) of particles having small pitch angles and partly by the greater enhancement of $v_\perp^2$ than $v_\parallel^2$ by the adiabatic acceleration accompanied by the injection (Cowley and Ashour-Abdalla, 1975). Hence the curvature drift term is sometimes dropped when the basic topology of the particle trajectory is examined in the magnetosphere.

In the inner magnetosphere the electric field of the magnetospheric origin, due both to the solar wind–magnetosphere coupling and the magnetotail reconnection, can be modeled roughly by a uniform field having the dawn-to-dusk direction. To this the electric field originating from the earth's rotation has to be added; since the rotation of the earth imposes corotational motion on the ionospheric plasma by viscous transport of momentum through the atmosphere, the ionosphere becomes electrically polarized in such a fashion as to oppose the electromotive force $(\boldsymbol{\Omega} \times \boldsymbol{r}) \times \boldsymbol{B}$ (where $\boldsymbol{\Omega}$ is the angular velocity of the earth's rotation), and the corotational component is induced in the motion of the magnetospheric plasma as the electric field due to this polarization is carried upward through the magnetosphere by the hydromagnetic action (Hines, 1964). Hence, neglecting the curvature drift and the time-dependent term, we can write the guiding center velocity in the inner magnetosphere as

$$\boldsymbol{u}_\perp = \frac{\boldsymbol{b}}{B} \times \boldsymbol{V}\Phi \qquad (92)$$

where

$$\Phi = -\frac{kR_{\mathrm{E}}^2}{r} - E_0 r \sin \psi + \frac{\mu M}{qr^3}. \qquad (93)$$

The first term of Φ represents the corotational electric field when $k = 14.5 \text{ mV/m}$ (or, 4.8×10^{-7} cgs) while the second term refers to the dawn-to-dusk electric field with strength E_0. r and ψ are radial distance and the azimuthal angle measured counterclockwise from the solar direction, respectively, and the magnetic field is represented by the earth's main field with the dipole moment M. Streamlines of the flow are given by $\Phi = \text{const}$. Note that the first and the third term of Φ represent circular motions around the earth, while the second term represents a uniform motion in the earth-sun direction.

Hence in the night-side magnetosphere an earthward injection of protons would occur when the westward electric field is intensified. Such an intensification of the large-scale electric field has indeed been detected by the whistler method at the onset of the expansion phase. (This method makes use of the character of the whistler atmospherics to propagate between northern and southern hemispheres along field-aligned ducts defined by plasma density irregularities and deduces the longitudinal component of the electric field from the radial motion of the ducts; the density irregularity is considered to drift with the velocity u_E, and its radial distance is derived from the nose frequency of the whistler dispersion curve.) Figure 78 gives an example of the electric-field observation during substorm. The left panel of this Figure shows, from the top, a component of IMF perpendicular to the dipole equator, the magnetospheric electric field at $L \sim 4$ derived from the whistler observation at Eights, Antarctica, and representative magnetograms from the night-side, auroral-zone stations. The magnetic midnight at Eights is indicated by a letter M below the electric-field data. The right panel contains low-latitude magnetograms in the corresponding interval. There is at least one case of a clear expansion phase during the interval, and its onset time is found to be ~ 0523 by these magnetograms as well as by the associated auroral observations. The Figure demonstrates that a large surging increase occurred in the westward component of the electric field in approximate time coincidence with this expansion-phase onset (Carpenter and Akasofu, 1972).

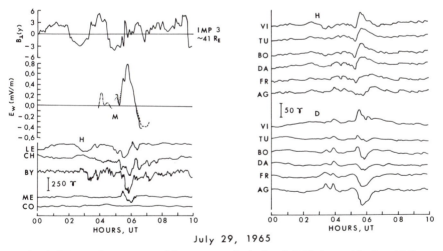

Fig. 78. Westward component of the magnetospheric electric field observed by the whistler method (left side, second panel). Onset of an expansion phase is signified at ~ 0523 UT by the low latitude positive bay seen in low-latitude H-component magnetograms (right side, upper panel) and the high-latitude negative bay seen in auroral-zone H-component records (left side, third panel) (Carpenter and Akasofu, 1972)

The geosynchronous orbit at $r = 6.6R_E$ provides a convenient monitoring spot of the injected plasma. The spectrogram (i.e., energy vs time curve) of Figure 79 represents a time history of the proton flux observed at this orbit by ATS 5. (The local magnetic midnight at this satellite is approximately 0700 UT.) Thick arrows below the spectrogram indicate the times when particles are injected; correction has been made of the time spent by the particles to drift to the satellite position after the injection has started. The lower panel of the Figure shows the ground magnetic

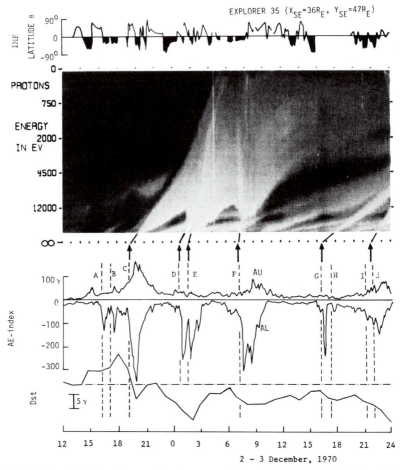

Fig. 79. Proton injection at expansion-phase onsets. *Top*: Latitude angle of IMF; *middle*: spectrogram of the proton flux measured by ATS 5; *bottom*: geomagnetic indices (Kamide and McIlwain, 1974)

activity by the AU, AL, and Dst[11] indices, and the onset times of substorm expansion phases are indicated by vertical dashed lines labeled A through J. It can be seen that except in weak substorms like events A and H the injection times agree with expansion-phase onsets within an accuracy of 10 min (Kamide and McIlwain, 1974). Thus energetic particles of the plasma sheet are injected toward the earth in consequence, possibly, of the occurrence of the reconnection at the X-type neutral line which is formed in the near-earth region of the magnetotail.

Trajectory of Injected Particles

Let us look more closely at trajectories of injected particles determined by Eqs. (92) and (93). The trajectory of the particle drift depends on the kinetic energy of the particle through the proportionality of the gradient-B drift velocity to the magnetic moment μ of the particle. Figure 80 illustrates the calculated streamlines of the proton drift in the equatorial plane; the streamlines are drawn for three representative values of μ for the assumed value of $E_0 = 0.36$ mV/m. The shaded area, called forbidden region, consists only of such streamlines that close in the vicinity of the earth and thus is inaccessible for particles from the plasma sheet. For protons having low enough μ that the ∇B drift is negligible as compared to the drift due to the earth's rotation, the forbidden region is produced as protons are deflected eastward by the corotation drift as they flow toward the earth (see top panel); for protons whose μ is high enough for the ∇B drift to dominate over the corotation drift the forbidden region results from the westward deflection of the earthward proton flow (see bottom panel). For protons in the intermediate range of μ the ∇B drift and the corotation drift tend to cancel each other, and the deepest penetration is expected to occur on the evening side at some specified value of μ that depends on the strength E_0 of the electric field (see middle panel) (Grebowsky and Chen, 1975).

The foregoing prediction of a simple model is basically consistent with particle observations during the main phase of a magnetic storm.

---

[11] The Dst index is essentially the average of the horizontal component observed at several low-latitude stations. Contributions from the geomagnetic main field and from the solar daily variation are subtracted, and the value is normalized to the equatorial value by assuming the disturbance field to be uniform and parallel to the dipole axis. For details, see Sugiura (1964) and Rostoker (1972).

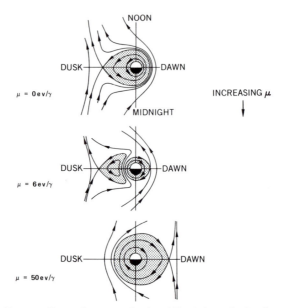

Fig. 80. Drift streamlines of protons having 90° pitch angle in the equatorial plane
(Grebowsky and Chen, 1975). μ: the magnetic moment of protons

Figure 81a illustrates the equatorial projection of the satellite (S³) orbit
when the observation was made; the magnetic storm started when the
satellite was at the position indicated as SSC, and its main phase (MP)
was in progress when the energetic protons were first detected in the
region indicated as nose event. The satellite crossed the plasmapause
(namely, the sharp change in the density of the ambient plasma of the
ionospheric origin, see Section IV.4) at the positions PP. In this instance
the depth of the penetration was the greatest for protons with $\mu \sim 40 \, \text{eV}/\gamma$
as demonstrated in Figure 81b where the phase space density of the
observed protons is plotted versus L values[12]. The decrease of penetration
depth at μ's above or below this critical value produces a characteristic
nose-like appearance in the energy-vs-time (namely, energy-vs-distance)
spectrogram as illustrated in Figure 81c (Smith and Hoffman, 1974).
Since the first adiabatic invariant is conserved, the 90°-pitch angle protons

---

[12] The L value is the equatorial crossing distance, in the unit of R_E, of the geomagnetic
main dipole field line that is equivalent to the *real* geomagnetic *main* field line. Here two
field lines are considered to be equivalent when a particle takes on identical values of the
first and the second adiabatic invariants on the two field lines. For details, see McIlwain
(1961).

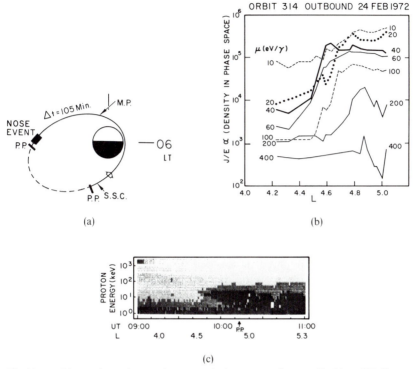

Fig. 81a–c. Observations of energetic protons during a magnetic storm (Smith and Hoffman, 1974). (a) Satellite orbit during the observation, (b) L-dependence of J/E (which is proportional to phase space density) of protons having magnetic moments μ of 10 to 400 eV$/\gamma$, and (c) the proton energy spectrogram versus distance

with $\mu \sim 40$ eV$/\gamma$ that constitute the nose structure can be traced back to the plasma-sheet protons having $E \sim 0.4$ keV in the 10γ field.

The modeling of the electric field by a steady, uniform field plus the corotation field is obviously a gross simplification, and a more refined model is needed to achieve better agreement with the observation. Regarding the spatial distribution, the dawn-to-dusk electric field seems to be enhanced in the post-2300 LT sector with a longitudinal extent of a few hours. This tendency has been recognized both from the local-time dependence of the particle arrival times observed at the geosynchronous altitude (Lezniak and Winckler, 1970) and from the confinement of the cross-L inward drift, deduced by the whistler method during substorms at $L \sim 4$, to a limited local-time sector (Carpenter et al., 1972). As for the time variation, the electric field is intensified and then reduced during a substorm. Since the distance that the particles can travel during the limited duration of the electric-field intensification is naturally limited, the penetration of protons to the inner magnetosphere is not likely to proceed as deeply as predicted by the steady-state model; indeed the nose structure has been encountered at L values that are higher than expected from the steady-state estimate (Ejiri et al., 1977).

Moreover, the unsteady nature of the electric field is essential to bring the injected particles to the trapped region. If both electric and magnetic fields were perfectly time-independent, the separation between the forbidden region and the streamlines from the tail would be absolute and the particles injected from the plasma sheet would all flow away to the magnetopause and be lost to the magnetosheath without entering the trapped orbit. In order for the injected plasma-sheet protons to be captured, it is necessary that the electric field weakens after the initial interval of the enhancement so that the forbidden region increases its dimension. The electric field during substorms indeed varies in the desired way, and a model study has been performed of the trapping process that takes place on the day side of the magnetosphere (Roederer and Hones, 1974). Further inward motion of the trapped particles has been considered to proceed by a diffusion process that is caused by the resonance between the azimuthal drift motion of particles and the time-variation of electric and/or magnetic perturbation fields (Schulz and Lanzerotti, 1974). The trapped protons are eventually lost by the charge exchange with cold protons of ionospheric origin (Smith et al., 1976).

Another point where the preceding model has to be considered oversimplified is the neglect of the polarization effect of injected particles. Due to the difference in trajectories between protons and electrons the polarization is produced and the electric field would be modified accordingly. This point will be taken up in Section IV.4.

Energy densities of protons and electrons in the earthward boundary regions of the plasma sheet are compared in Figure 82. Curves labeled "proton" and "electron" represent energy densities of protons between 90 eV and 48 keV and of electrons between 80 eV and 46 keV, respectively. As the figure shows, the electron energy density is almost an order of magnitude lower than the proton energy density. In consequence of this difference electrons injected from the plasma sheet do not carry as much energy as protons do, and their contribution to the kinetic energy content of the inner magnetosphere can usually be ignored. [The "trapping boundary" and "plasmapause" in the figure indicate crossings of the outward boundaries of energetic trapped electrons ($E > 40$ keV) and of cold, ambient plasma, respectively.]

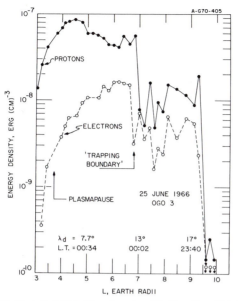

Fig. 82. Electron (80 eV $\leq E \leq$ 46 keV) and proton (90 eV $\leq E \leq$ 48 keV) energy densities near local midnight (Frank, 1971). λ_d and L.T. refer to dipole magnetic latitude and geomagnetic local time

IV.3 Geomagnetic Effect of the Injected Particles

Particles injected and trapped in the inner magnetosphere distort the geomagnetic field by the current associated with their drift motion. The development of this perturbation field reflects the condition of IMF which controls the energy supply to the magnetosphere.

Structure of the Ring Current

When particles are distributed axisymmetrically inside the magnetosphere, the current carried thereby forms a closed ring and the resulting distortion of the geomagnetic field can be calculated using Eq. (11).

The calculation requires a method of successive approximation, since the current density j_P given by Eq. (11) depends upon the magnetic field it produces. The algorithm has been developed by Hoffman and Bracken (1967) and others, and Figure 83 is a comparison of the calculated and the observed ΔB during the June 17–20, 1972 storm (Berko et al., 1975).

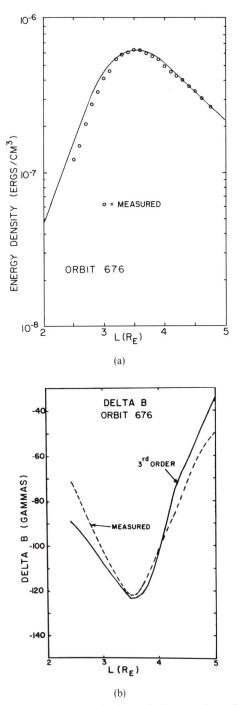

Fig. 83a,b. (a) Observed proton energy density, and (b) comparison of the magnetic-field perturbation calculated from the above (solid curve) with the one observed simultaneously (dashed curve) (Berko et al., 1975)

The proton energy density shown by a solid curve in Figure 83a represents a smooth fit to the measured values integrated over energy (1 to 872 keV) and pitch angle. The observation was made during the recovery phase of this storm when the distribution of protons around the earth had become axisymmetric and the Dst index was about -80γ. The solid curve of Figure 83b is the magnetic perturbation field that is calculated from the observed energy and pitch angle distributions, and it agrees very well with the measured value of the perturbation field shown by a dashed curve. Thus the idea that the ring current is carried by the energetic protons introduced deep into the magnetosphere is well substantiated by the observation.

The meridional cross section of the ring current and the resulting distortion of the geomagnetic lines of force calculated by Hoffman and Bracken (1967) are illustrated in Figure 84. (The proton energy distribution on which these Figures are based is not identical to the one shown in Figure 83a but is quite similar.) Figure 84a shows the distribution of the azimuthal current density where positive and negative values indicate eastward and westward currents, respectively. The current is directed westward beyond $L \sim 3.5$ where the pressure gradient is directed inward, whereas inside $L \sim 3.5$ the current flows eastward due to the outward pressure gradient. The perturbation field that results from this current system acts to reduce the low-latitude geomagnetic field on the ground as the effect of the westward current dominates. The field line traces shown in Figure 84b demonstrate that the field lines are not immensely distorted even when the proton energy content is set at a high level that corresponds to $\beta \sim 3$ at the heart ($L \sim 4$) of the proton belt.

The nature of the j_P current can be related to the guiding center velocity when it is written as

$$j_P = -V \times \left(\frac{P_\perp}{B^2} B\right) + \frac{P_\perp}{B^2} b \times VB + \frac{P_\parallel}{B} b \times (b \cdot V)b \qquad (94)$$

where P represents the sum of proton and electron pressures and b is the unit vector in the direction of B. [The identity $(b \cdot V)b + b \times (V \times b) = 0$. which follows $|b|^2 = 1$, has been used to derive the above.] In this expression the second and the third terms represent the effect of the gradient-B drift and of the curvature drift, respectively, and both cause decreases in the low-latitude ground field since they are directed westward in the geomagnetic field. The first term, often referred to as the magnetization current, originates from the magnetic dipole moment associated with the gyrating motion of charged particles in the geomagnetic field, and it acts to reduce the field only in the region where the particles are locally present.

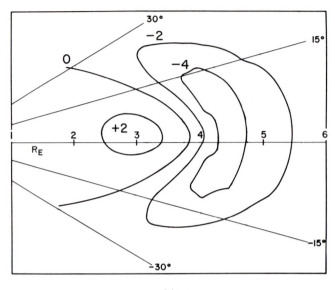

(a)

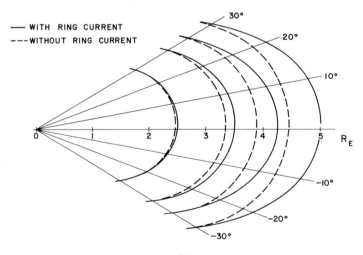

(b)

Fig. 84a,b. (a) Meridional cross section of the calculated ring current profile and (b) calculated field line distortions (Hoffman and Bracken, 1967). The current is in the unit of 3×10^{-9} A/m^2

The current given by this term is the principal source of the field distortion observed in the heart of the proton belt. However, its ground effect, which is the increase in the low-latitude field, is insignificant relative to the effects of the two other terms.

During the main phase the depression observed in the horizontal component of the ground magnetic field in low latitudes is not axially symmetric, the depression being deeper on the afternoon side than on the morning side (Meng and Akasofu, 1967). An example of this feature is shown in Figure 85a where late afternoon hours for each station (1800 to 2400 LT) are indicated by heavy bars for each trace. Sometimes the asymmetry is so pronounced that the depression is almost absent on the morning side while a well-developed depression is observed on the afternoon side; an example can be found around 0800 UT, June 16. From the ground observation alone it is not possible to tell whether the origin of this asymmetry is in the structure of the magnetospheric current system or in the local-time dependence of the ionospheric current flow that might be produced at the same time, but the Explorer 26 traversals of the ring current region during the same interval confirmed that there was asymmetry in the ring current intensity. Figure 85b shows the magnetic field observed at L of $4 \sim 5$ as the satellite cruised from the morning to the afternoon sector across the noon meridian (see LT given at the bottom). As seen in the top panel, the depression of the observed field strength from the main field is much greater on the afternoon side (~ 1400 LT) than on the morning side (~ 0930 LT) (Cahill, 1970). Furthermore the OGO 2 observations of the perturbation field in the height range of 400 to 1500 km have clearly demonstrated that the asymmetry has a magnetospheric origin; essentially the same asymmetry as noted on the ground was detected simultaneously above the ionosphere (Langel and Sweeney, 1971).

The main phase is an interval during which substorms are activated. This is demonstrated in Figure 86 where it is seen that the intensity of the auroral electrojet (represented by the AL index) is high when DST = $|\text{Dst}|$ is increasing. Hence it would be natural to attribute the asymmetry in the low-latitude field depression to the progress of the particle injection from the magnetotail. Until the injected protons enter the trapped orbit and encircle the earth, the ring current is not closed by itself and forms a partial ring current, as illustrated in Figure 68. The circuit is closed via the ionospheric path (see Section IV.4).

Thus the DR field observed on the ground during the early stage of the ring current formation is not simply due to j_P, and model calculations of

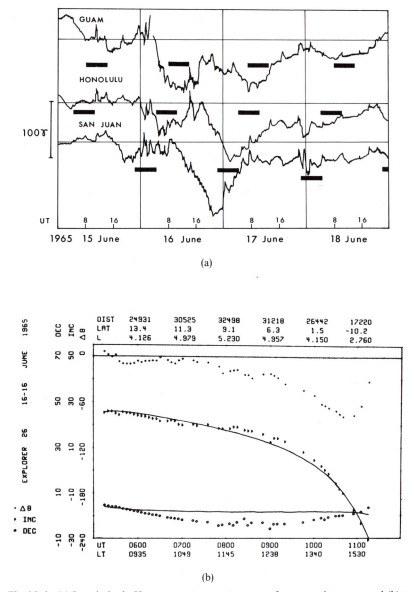

Fig. 85a,b. (a) Low-latitude H-component magnetograms of a magnetic storm, and (b) a satellite magnetometer record obtained in the early stage (0500 ~ 1200 June 16) of the above storm (Cahill, 1970). In (b) the three plots represent, from the top, strength of the magnetic perturbation field, inclination, and declination. The solid curves represent the prediction by Jensen-Cain's model of the earth's main field

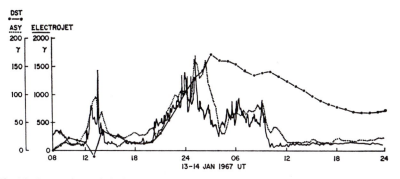

Fig. 86. Comparison of the intensity of the westward electrojet, range of the ring-current asymmetry (ASY), and the Dst index during a storm (Kawasaki and Akasofu, 1971b)

the DR field have been performed under various conditions of the longitudinal extent of the partial ring current and of the current closure in the ionosphere. According to these model calculations the asymmetric DR field observed on the ground is mainly due to the field-aligned portion of this circuit; the perturbation field produced on the ground by the 'partial ring' portion of the circuit does not vary much with local time since the partial ring is located more than a few earth radii away from the center of the earth.

There is a problem, however, regarding the local time position of this current circuit. Since the asymmetric DR field observed on the ground is most intense around the 1800 LT meridian, the dusk sector was suggested as the seat of the circuit when the idea was originally introduced (Akasofu and Chapman, 1964; Kamide and Fukushima, 1971). According to this view there should be a reversal in the polarity of the field-aligned current around the dusk meridian; the current should flow into the ionosphere before dusk and out of it after dusk. When the global distribution of the field-aligned current finally became mapped, however, the observation did not agree with the foregoing expectation. As seen in Figure 71a the reversal in the polarity of the field-aligned current occurs around midnight rather than around dusk. (The partial ring current system of Figure 68a has in fact been drawn centered around midnight so that it is not actually consistent with the observation of the asymmetric DR on the ground.) Thus, although the observation of the nose structure and the associated depression of the local magnetic field supports the view that the magnetosphere is more inflated in the dusk sector (Lee and Cahill, 1975), we have yet to find out how that feature is related to the local-time asymmetry of the DR field observed on the ground.

The asymmetry in the DR field is reduced and the ring current tends toward being symmetric when the storm enters the recovery phase after the peak in |Dst| is reached (Kawasaki and Akasofu, 1971b; Kamide and Fukushima, 1971). This feature is seen in Figure 86 where DST = |Dst| is compared with ASY, which is defined as the range between the upper and lower envelopes of the H-component records, the Sq variation having been removed, of representative low-latitude stations. It is seen that the DR field was highly asymmetric (DST $\simeq$ ASY) when it was growing in association with the high electrojet activity after ~ 2000, January 13 but became symmetric after the electrojet activity ceased. In terms of the dynamics of the injected protons, the symmetrization of the ring current is thought to occur when the particle injection has terminated and the protons trapped due to the weakening of the electric field have encircled the earth. (Another feature recorded in Figure 86 during $1200 \sim 1400$, January 13 does not serve to illustrate the DR-field development, because it is dominated by large SIs.)

Development and Decay of Ring Current

Since the injection of energetic protons into the inner magnetosphere represents one of the consequences of the energy transfer from the solar wind to the earth's magnetosphere, the growth rate of the ring current should depend on the reconnection rate at the day-side magnetopause. Hence it is expected that the growth rate depends on the IMF condition in a way similar to the activity of the magnetic substorm. The decay rate of the ring current, on the other hand, is determined by the rate at which injected protons are lost mainly by the charge exchange, and hence it is expected to be proportional to the strength of the ring current. These expectations are indeed borne out by observations. Figure 87a is a comparison of the decay rate of I_{DR} to I_{DR} itself, based on observations during intervals when B_z of IMF is directed northward so that (as shown immediately below) the deep injection would not be in progress. Here I_{DR}, the local-time averaged DR-field strength on the ground, is derived from the Dst index by

$$I_{DR} = \text{Dst} - b(P_d)^{1/2} + c \qquad (95)$$

where the second term subtracts from Dst the contribution from the magnetopause current produced by the dynamic pressure P_d of the solar wind, and the third term represents the quiet-day constant. From Figure 87a the reciprocal of the decay time is found to be roughly $-3.6 \times 10^{-5}/\text{s}$. Figure 87b is a comparison of the growth rate, or injection rate, of I_{DR}

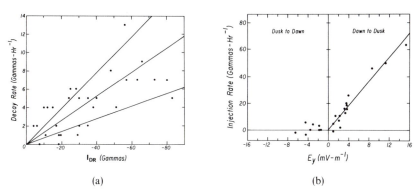

(a)   (b)

Fig. 87a,b. (a) Decay rate of the DR-field intensity as a function of the DR-intensity, and (b) the growth rate of the DR-field as a function of the dawn-to-dusk component of IMF (Burton et al., 1975)

to the dawn–dusk component E_y of the interplanetary electric field, which is the product of the solar wind velocity and the southward component of IMF; intervals have been chosen in which P_d was constant and a time delay of 25 min has been assumed between the two variables. It is seen that the growth rate is nearly proportional to E_y when the latter is positive, namely, when B_z of IMF has a southward polarity, but there is very little injection when B_z is northward. Thus the development of the ring current depends on the southward polarity of IMF more critically than does the activity of magnetic substorms.

On the basis of these results and earlier estimates of characteristic rate coefficients [including the estimate of the solar wind pressure effect given by Eq. (42)], Burton et al. (1975) attempted to predict the strength of the ring current by a simple empirical formula:

$$\frac{dI_{DR}}{dt} = F(E_y) - aI_{DR} \tag{96}$$

where

$$F(E_y) = 0 \qquad\qquad \text{if } E_y < +0.50 \text{ mV/m}$$

$$F(E_y) = d(E_y - 0.5) \qquad \text{if } E_y > +0.50 \text{ mV/m}$$

with

$$a = 3.6 \times 10^{-5}/\text{s} \qquad b = 0.20\gamma/(\text{eV cm}^{-3})^{1/2}$$

$$c = 20\gamma \qquad\qquad d = -1.5 \times 10^{-3}\gamma/(\text{mVm}^{-1}) \text{ s}.$$

Since rapidly oscillating interplanetary electric fields have been found to be not very efficient in causing the growth of the ring current, E_y has been low-pass filtered with a corner frequency of 2 cph and an attenuation of 6 dB per octave. From I_{DR} thus obtained, Dst is calculated using Eq. (95), and the result has been compared with the observed Dst. As illustrated by two examples in Figure 88, the predicted and the observed Dsts agree extremely well.

In the case examined in Figure 88a the main phase was initiated when E_y became large and positive. It was preceded by an initial phase of increased positive Dst since the solar-wind dynamic pressure was enhanced

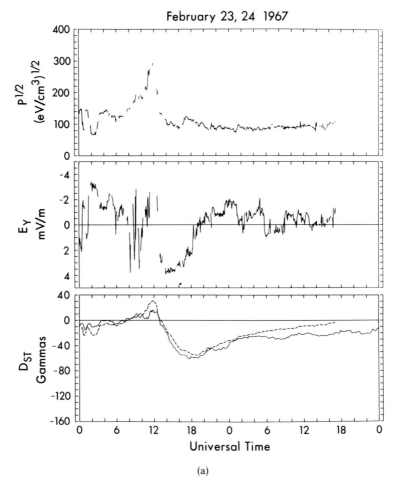

(a)

Fig. 88a,b. A couple of comparison between the predicted and the observed Dst. Each column contains, from the top, the square root of the solar-wind dynamic pressure, the dawn-to-dusk component of the interplanetary electric field, and the predicted (dashed curve) and the observed (solid curve) Dsts (Burton et al., 1975)

prior to the main phase. The recovery phase was started when the injection rate dropped below the decay rate as a result of the decrease in E_y. Since increases in the dynamic pressure of the solar wind and in the dawn–dusk component of the interplanetary electric field are basically independent phenomena, the sudden commencement and the initial phase do not necessarily take place prior to the main phase, as typified by the example in Figure 88b. Similarly, 'sudden commencements' can occur without being followed by the main phase, and these cases have been known as sudden impulses. Nevertheless, there is a high possibility that large, positive SIs are followed by large DR fields; this is because both

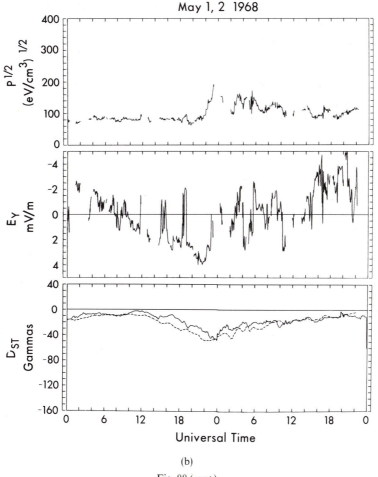

(b)

Fig. 88 (cont.)

the solar-wind velocity and the IMF strength tend to be augmented during periods of enhanced solar-wind pressure, and there is a good chance that large E_y is produced when IMF turns southward. This appears to be the reason why the elements comprising the 'magnetic storm' have often been observed as a set and the concept of the magnetic storm was coined (Burton et al., 1975).

According to Siscoe and Crooker (1974b), the growth rate of $-1.5 \times 10^{-3}\gamma/(\mathrm{mVm}^{-1})$ s derived from Figure 87b seems to be reasonably consistent with the current understanding of the energy transfer process from the solar wind to the radiation belt. The energy of the solar wind may be considered to be transferred to the interior of the magnetosphere by the Lorentz force exerted on the connected field lines. The force can be expressed as a tangential stress equal in magnitude to $B_n B_t/\mu_0$, where B_n is the component of the field normal to the boundary and B_t is the tangential component. The total force acting on a length $L_T R_E$ of the tail is $F_T = 2\pi R_T L_T R_E B_n B_t/\mu_0$, where R_T is the tail radius. The power P transferred to the field is then $F_T V$, where V is the speed of the solar wind adjacent to the tail. Hence

$$P = 2\pi R_T L_T R_E V B_n B_t/\mu_0. \tag{97}$$

Using the conservation of the magnetic flux, the field B_n emerging from the tail surface can be related to IMF roughly by

$$\pi R_T B_n V = |B_z| V L_M R_E$$
$$= E L_M R_E \tag{98}$$

where $L_M R_E$ is the longitudinal extent of the upstream IMF field lines that are to be reconnected, and E represents the interplanetary electric field. For the sake of simplicity, only the southward component B_z of IMF is considered, and the spatial variation of V is ignored. Combining Eqs. (97) and (98), the rate of energy input can be expressed as

$$P = (2R_E^2 L_T L_M B_t/\mu_0)E. \tag{99}$$

The rate of growth of the kinetic energy content of the magnetosphere, on the other hand, can be estimated from Eq. (89) by

$$\frac{dU_T}{dt} = \frac{3}{2}\frac{U_D}{B_D}\frac{d|I_{DR}|}{dt} \tag{100}$$

where

$$U_D = \frac{4\pi}{3\mu_0} R_E^3 B_D^2.$$

It would be reasonable to assume that $dU_T/dt = P/\eta$, where the factor $\eta > 1$ is to account for other sinks of energy in the magnetosphere such as Ohmic dissipation in the ionosphere and particle precipitation into the atmosphere. Consequently, we find

$$\frac{d|I_{DR}|}{dt} = \frac{B_t L_M L_T}{\pi \eta B_D R_E} E \qquad (101)$$

where $\eta = 2$ and $B_t = 30\gamma$ would be appropriate. Since $B_D = 3 \times 10^4 \gamma$, the numerical value of the factor to E is $3 \times 10^{-5} L_M L_T$ if $|I_{DR}|$ is expressed in the unit of γ and E in the unit of mV/m. A reasonable choice of L_T would be ~ 15, which represents the distance to the near-earth neutral line, since the energy that is supplied beyond this distance would be swept away, instead of being fed to the inner magnetosphere, when the near-earth reconnection operates. Then the factor 1.5×10^{-3} obtains, if we take L_M (the longitudinal extent of the reconnected field line) to be about 3, which appears to be consistent with the value of $L_M \sim 10$ obtained by Gonzalez and Mozer (1974) in view of the approximate nature both of the above estimate and of the Gonzalez-Mozer model.

During intervals of northward IMF, weak substorms are observed, but little injection of energy to the ring current is recognized. The reason for this probably lies in the weakness of the dawn-to-dusk electric field that is associated with weak substorms. If the electric field is weak, particles injected from the tail cannot reach low L shells, and the efficiency of the particle replenishment to the trapped region is very much reduced even if the injection may very well occur.

IV.4 Alfven Layer and the Plasmapause

Alfven Layer

On account of the difference between trajectories of protons and electrons expressed by Eq. (91), the plasma injected from the tail becomes electrically polarized. This effect would be most significant at boundaries of the injected plasma where the density gradient is high; even when numbers of protons and electrons were made perfectly equal by some hypothetical mechanism at one instant, the neutrality breaks down immediately as protons and electrons drift apart due to the differential motion. (Formally, the effect is expressed by the first term of Eq. (76).) Thus a layer having a net positive or negative charge is produced at the surface of the injected plasma. This layer has been designated as the Alfven layer after Alfven (1955) who first suggested its formation. The resulting electric field drives a three dimensional circuit as represented in Figure 68 by current a′, b′, c′ and d′.

The development of the Alfven layer and the associated modification of the electric field have been pursued using numerical methods. According to Jaggi and Wolf (1973), the problem is formulated as follows. In the magnetotail as well as inside the polar cap, a dawn-to-dusk electric field of a fixed strength is assumed to exist. The electric field causes the inner edge of the plasma-sheet ions to move with a velocity $u_\perp$ given by Eq. (92), where Φ is given by Eq. (93) plus the ∇B drift terms originating from additional magnetic fields due to magnetopause and magnetotail currents. For simplicity all the particles of the plasma sheet origin are assumed to drift in the equatorial plane, and the feedback effect of the particle motion on the magnetic field is ignored. Also, the plasma sheet electrons are assumed to be much colder than their proton counterparts so that the electron motion is approximated by the electric-field drifts alone. Then the net charge that is produced per unit time per unit length dl of the Alfven layer is given by the flux of the gradient-B drift motion of protons across dl;

$$dJ = \frac{n\mu}{B} \frac{dB}{dl} dl \tag{102}$$

where n is the ion density integrated along the field line and dB/dl is the change, per unit length eastward along the Alfven layer, of the magnetic field. The electric-field drift does not contribute to the polarization since both protons and electrons share this component of the drift velocity.

The coupling with the ionosphere acts to suppress the growth of the polarization as the accumulated charge is discharged by the ionospheric current. Since the ionospheric current is divergent or convergent in the region where dJ flows in or out, there has to be a jump $[E_n]$ in the component of the electric field perpendicular to the ionospheric projection of the Alfven layer. $[E_n]$ is given by

$$\Sigma_P [E_n] = \frac{s}{2} \frac{dJ}{dl} \tag{103}$$

where s is the scaling factor and the current dJ is assumed to be split equally between northern and southern ionospheres. The electric field is then solved by using both Eq. (103) and the condition at the edge of the polar cap (inside which uniform dawn-to-dusk field is assumed) as boundary conditions. To avoid complexities at the equator the latitudinal current is set equal to 0 at the latitude of 21.2°.

Figure 89a illustrates the electric potential distribution that would be obtained in the Jaggi-Wolf model if the polarization in the Alfven layer were absent. The polar-cap electric field is assigned by the potential distribution on the curve I, and the nonuniformity of the electric field earthward of this curve reflects the polarization produced in the ionosphere by the nonuniformity of the ionospheric conductivity. In contrast, Figure 89b shows the potential distribution that is obtained when the formation of the Alfven layer is taken into account. To yield Figure 89b the ion sheet consisting of protons with $\mu = 50 \, \text{eV}/\gamma$ and density, per unit magnetic flux, of 6.3×10^{20} ions/Wb (which corresponds to $n \sim 1/\text{cm}^3$ at $L = 7$) is set to move from the initial boundary position at $|x| \gtrsim 16R_\text{E}$, and its motion is followed for 6 h under the electric field modified by polarizations that are created both in the magnetosphere (at the Alfven layer) and in the ionosphere (due to local-time and latitude dependences of Σ_p). Field lines are assumed to be equipotentials throughout. The principal difference between Figure 89b and Figure 89a lies in the weakening of the electric field earthward of the Alfven layer. The weakening results as the polarization electric field produced by the differential proton/electron drift motions is directed dusk-to-dawn, namely, opposite to the dawn-to-dusk electric field applied from outside. The reduction in the field strength is particularly pronounced on the night side where the Alfven-layer polarization cannot be neutralized efficiently due to the low conductivity in the ionosphere, and the relaxation time of the electric field has been estimated to be ~ 3 min for the night side while it is ~ 5 h for the day side (Jaggi and Wolf, 1973). If the applied electric field is entirely time-independent, it is conceivable that the electric field from outside is shielded completely at the Alfven layer (Block, 1966). (Since the purpose of the calculation is to examine the modification of the dawn-to-dusk electric field applied by outside sources, the corotational electric field is not included in the potential distributions displayed in Figure 89a and b. To obtain complete potential contours in the nonrotating frame of reference, the potential of the corotational field has to be added to these.)

The potential distribution at ionospheric heights is shown in Figure 90. The Figure is obtained by assuming the ion magnetic moment to be slightly higher ($\mu = 200 \, \text{eV}/\gamma$) than the one used in Figure 89, and it illustrates the equipotential contours at 7 h after the ion sheet started to move. The inner circle represents the edge of the polar cap whereas the dash-dot curve corresponds to the ionospheric projection of the Alfven layer. Because many hours have passed since the start of the ion-sheet motion

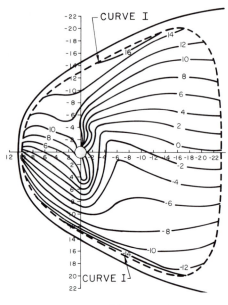

(a)

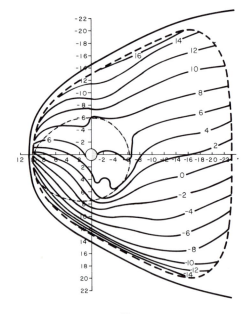

(b)

Fig. 89a,b. (a) Convection potential computed assuming no Birkeland current from the Alfvén layer. The numbers on the curves are potentials in kV. (b) The potential 6 h after the ion sheet started in from the tail. Dash-dot curve represents the position of the Alfvén layer (Jaggi and Wolf, 1973)

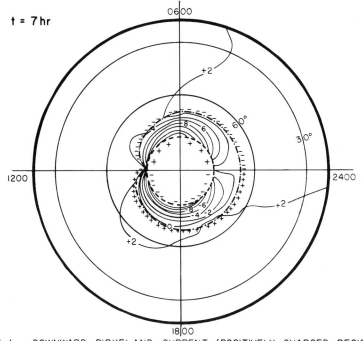

t = 7 hr

+ + + DOWNWARD BIRKELAND CURRENT (POSITIVELY CHARGED REGION)
– – – UPWARD BIRKELAND CURRENT (NEGATIVELY CHARGED REGION)

Fig. 90. Potential distribution 7 h after the ion sheet started in from the tail, shown in the ionosphere (Wolf, 1974)

from the tail, the electric field has become confined almost perfectly to the region poleward of the Alfven layer. The predicted polarity distribution of the field-aligned current at the Alfven layer, namely downward on the evening side and upward on the morning side, is in reasonable agreement with the observed one shown in Figure 71a, but the distribution seems to depend on the assumed value of μ (Jaggi and Wolf, 1973). Moreover, the available model has not been able to reproduce the proton nose structure displayed in Figure 81c. Together with the explanation of the local-time position of the asymmetric DR field (discussed in Section IV. 3), dynamic behavior of the injected particles remains to be pursued.

Plasmapause

So far we have been concerned only with the more energetic of the plasma populating the magnetosphere, namely, the particles that reside in the tail plasma sheet and occasionally flow earthward to inflate the inner magnetosphere. The average energy of these particles is in the range of 0.1 to 10 keV in the magnetotail and is increased by adiabatic acceleration processes as they move earthward. In addition to these particles, the magnetospheric plasma comprises another component that originates from the ionosphere. The average energy of this component is low—in the range of 0.1 to 1 eV only, but in number the cold plasma exceeds the energetic population, and it represents the principal constituent of the inner magnetosphere. The cold plasma is produced in the ionosphere by the ionizing effects of the solar radiation and the precipitating energetic particles, and it is then distributed throughout the magnetosphere by the pressure gradient and the centrifugal force that act against the gravitational force.

In the region of closed field lines, the supply of the cold plasma from the ionosphere to the magnetosphere continues until the state of the hydrostatic equilibrium is reached, namely, until the following equations become satisfied.

$$\frac{d(n_e T_e)}{ds} = -\frac{n_e m_+}{2k} f \tag{104}$$

$$\frac{d(n_i T_i)}{ds} = -\frac{n_i \{m_i - (m_+/2)\}}{k} f \tag{105}$$

where the distance ds is measured along the line of force and m_+ is the temperature-weighted ion mass average given by

$$m_+ = \frac{2 \sum_i m_i n_i T_e(s)/T_i(s)}{\sum_i n_i (1 + T_e(s)/T_i(s))}$$

and

$$f = (g - \Omega^2 \rho) \cdot b$$

where g refers to the gravitational force, Ω refers to the angular velocity of the earth's rotation, ρ refers to the distance from the rotation axis, b refers to the unit vector in the direction of B, and other variables have usual meanings (Angerami and Thomas, 1964). Eqs. (104) and (105) essentially represent the balance among the gravitational force, centrifugal force, pressure gradient, and the electric field due to the vertical separation of ions and electrons caused by the difference in their masses. The boundary condition is represented by electron and ion densities in the ionosphere.

At $L \sim 4$ it takes about a week to fill the closed field line that was emptied (Park, 1970).

The tendency toward the hydrostatic equilibrium, however, is disturbed as the plasma in the magnetosphere participates in the convective motion across field lines. When the sunward convection (due to dawn-to-dusk electric field) dominates the corotational motion in the outer magnetosphere, the ionospheric plasma that is supplied along field lines to the outer magnetosphere is transported to the sunward magnetopause and is lost to the magnetosheath. The flow paths of the low-energy plasma, for which the VB- and curvature-drifts can be neglected relative to the electric field drift, are illustrated in Figure 91 assuming that the uniform dawn-to-dusk electric field of 0.3 mV/m is applied from outside. Figure 91a shows the case when there is little attenuation of the applied electric field at the Alfven layer, and the streamlines represent equipotentials of the corotational plus dawn-to-dusk electric field everywhere. (The figure is essentially identical to the first panel of Figure 80.) Figure 91b represents the case when there is perfect shielding of the applied electric field at the Alfven layer, which is approximated by a circle of $4R_E$ radius. In both cases, the plasma supplied to the region lying outside the solid curve

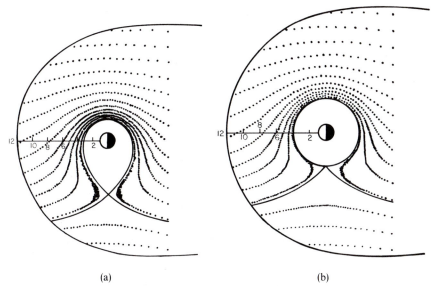

(a)   (b)

Fig. 91a,b. Convection streamlines, including the corotational motion, of cold particles in the equatorial plane of the magnetosphere (a) when the shielding effect of the Alfven layer is negligible and (b) when there is perfect shielding at the Alfven layer (Kavanagh et al., 1968). The distance between successive dots is the distance which a particle travels in 10 min

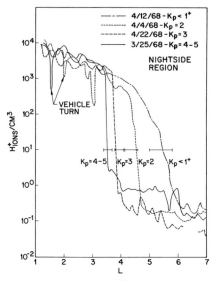

Fig. 92. Examples of plasmapause crossings at different levels of magnetic activity. The data were obtained by OGO 5 between midnight and 0400 LT (Chappell, 1972)

encircling the earth is lost to the magnetosheath in about a day or less, namely, within times that are appreciably shorter than the time scale of several days required to fill the magnetic tube of force to the level of the hydrostatic equilibrium. Inside the solid curve where the corotational motion is dominant, on the other hand, the presence of the flow does not cause loss of plasma. Hence a sharp gradient of the cold plasma density is expected to be formed inside the magnetosphere at the position of the solid curve.

Such a density gradient has indeed been detected. The earliest detection of this structure, named plasmapause, was made by the whistler method of deducing the magnetospheric electron density (Carpenter, 1966), and ample confirmation has since been offered by satellite observations (Chappell, 1972). Figure 92 is a proton density vs. L plot that demonstrates the presence of the sharp density gradient inside the magnetosphere. Characteristically the distance to the plasmapause varies with the geo-magnetic activity and it tends to be smaller when Kp is higher. (The Kp used is the average in the 6-h period preceding the observation.) Figure 93 shows the equatorial section of the average plasmapause derived from more than 150 OGO 5 plasmapause crossings, and clearly there is a local time dependence in the plasmapause position.

The question then is whether the plasmapause is due to the magneto-spheric flow pattern of type Figure 91a or b. Nishida (1966c) proposed the a-type model on the basis of the observation that DP2 fluctuations,

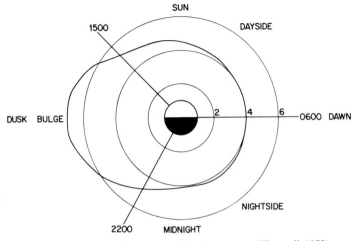

Fig. 93. Average configuration of the plasmapause (Chappell, 1972)

which are correlated with IMF variations, appear at the equatorial daytime stations and thus testify to the penetration of the external electric field deep into the magnetosphere, while the b-type model was favored by Block (1966). Subsequently a number of observations rendered support to the thesis that the electric field of external origin does exist inside the plasmasphere. The most direct piece of evidence is provided by the electric field observation by the whistler method; the electric-field variations shown in Figure 78 were recorded inside the plasmasphere. Moreover, the proton nose event presented in Figure 81 was detected inside the plasmasphere, whereas high energy protons from the tail would have been deflected before reaching the plasmapause if the b-type flow pattern had prevailed. The absence of the complete shielding at the Alfven layer is indeed to be expected since the electric fields that arise from the IMF effect and the magnetotail dynamics are by no means steady and tend to vary with a characteristic time scale of about 1 h. Although the time variations of the electric field introduce some complexities that will be discussed later, the plasmapause produced by the 'gusty' convection has been shown to be essentially of the same shape as the a-type boundary formed by the steady convection (Chen and Wolf, 1972). The local time dependence of the plasmapause configuration, characterized by the presence of the bulge in the evening sector (Fig. 93), compares nicely with the expection of the Figure 91a model too.

In the preceding model, the dusk (1800 LT) section of the plasmapause is located at the distance where the eastward corotational velocity and the westward convection velocity applied from outside are equal and opposite. Using Eqs. (92) and (93), this condition can be expressed at the equator by

$$\frac{kR_E^2}{r^2} - E_0 = 0. \tag{106}$$

Hence the distance to the dusk plasmapause is

$$r = 3.81 R_E / \sqrt{E_0}$$

where E_0 is the strength of the convection field in mV/m; the plasmapause shrinks as E_0 is increased. Since the Kp index can be considered to present a rough measure of the strength of the magnetospheric electric field, the decrease of the plasmapause distance with an increase in Kp noted in Figure 92 is also in agreement with the model.

Recent efforts to explain the dynamic behavior of the plasmapause by the convection model have developed more sophisticated expressions of the convection electric field, for example

$$\Phi = -\frac{kR_E^2}{r} - A \left(\frac{r}{R_E} \right)^2 \sin \psi \tag{107}$$

where the r^2 dependence of the potential of the convection electric field can be interpreted to reflect the partial shielding of the external electric field by the Alfven layer. The factor A determined empirically depends on Kp as

$$A = 0.045(1 - 0.159\ \text{Kp} + 0.0093\ \text{Kp}^2)^{-3}\ (\text{kV}/R_E^2).$$

Figure 94 is a comparison of the observed plasmapause positions with the prediction by the model. In this model the plasma that has been on closed field lines for 6 days or more is tentatively identified to constitute the plasmasphere, and the region occupied by such plasma is shaded. The shaded area agrees fairly well with the heavy orbit trace, which denotes regions where the high cold plasma density was indicated by the observation (Maynard and Chen, 1975). The plasma tail that extends from the main body of the plasmasphere is the product of the change in the convection electric field. It is formed as the plasma is peeled away and transported sunward by the enhanced convection (see middle panel). When the intense convection subsides the plasma tail wraps itself around the main body of the plasmasphere due to the corotational effect that has become dominant, as illustrated by the example in the left panel.

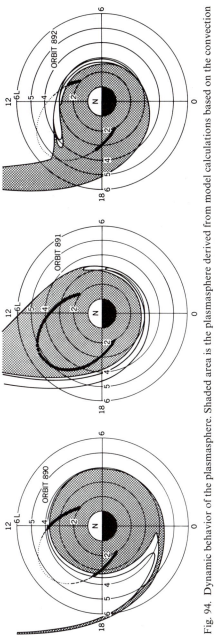

Fig. 94. Dynamic behavior of the plasmasphere. Shaded area is the plasmasphere derived from model calculations based on the convection theory, and heavy orbit trace denotes regions of higher plasma density (Maynard and Chen, 1975)

As will be discussed in Chapter V the growth rate of ion cyclotron instability, which causes the pitch-angle scattering of energetic protons and activates precipitation of energetic protons along geomagnetic lines of force, is inversely proportional to the total plasma density. Hence during the recovery phase the deposit of energy in the ionosphere occurs in the region that received injected protons during the disturbance (since the forbidden region shrank under the enhanced convection electric field) but becomes enclosed by the plasmapause once the disturbance has subsided (since the dimension of the plasmapause is increased due to the reduced convection electric field). One of the manifestations of the energy deposited in the ionosphere is the enhancement of the 630.0 nm emission called the stable auroral red (SAR) arc that appears at L of 2 to 4, and a correlation has been noted between the SAR-arc intensity and the Dst index (Rees and Roble, 1975). The ion cyclotron waves that are excited by the instability are observed on the ground as pulsations of the pc1 category (see Section V.4).

IV.5 Sq, Geomagnetic Solar Daily Variation

On geomagnetically quiet days when the magnetospheric and the ionospheric currents driven by the energy input from the solar wind are either very weak or kept at fixed levels, the geomagnetic field on the ground is dominated by a regular variation that has a periodicity of one solar day. This variation, called Sq, has been considered to originate from the dynamo action of the wind system in the ionosphere, and hence its interpretation requires an understanding of the atmospheric dynamics. Since the physics of the earth's atmosphere is an extensive subject that cannot be covered in full by this short monograph, we shall discuss only some basic features of Sq and related phenomena.

Sq and Ionospheric Dynamo

The equivalent ionospheric current system of Sq is given in Figure 95. The current system is drawn by taking the level of the quiet-day daily mean as a baseline level, and coordinates are dip latitude and solar local time. The contribution from the current induced within the earth has been excluded by the spherical harmonic analysis. Figure 95a is the view from the magnetic equatorial plane at 1200 LT, and Figure 95b is the view from over the north magnetic pole. (The preceding current system is based on observations at dip latitudes of less than 61° only;

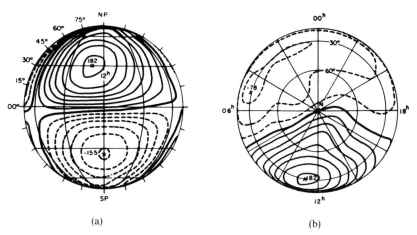

(a)   (b)

Fig. 95a,b. Average equivalent current system of Sq during the IGY (external part only). (a) View from the noon side and (b) view from over the north pole (Matsushita, 1965). The current between adjacent streamlines is 2.5×10^4 Amp. The thick solid curves indicate the zero-intensity lines, and thin solid and broken curves show counterclockwise and clockwise current vortices, respectively. Numbers given at the center of vortices indicate the total current intensities of these vortices in units of 10^3 Amp

the current drawn across the polar cap is the extrapolation from the low-latitude equivalent currents.) The principal features of the Sq current system are the current vortices in the dayside hemisphere that are centered around the noon meridian at dip latitudes of about $\pm 30°$. The current flows westward on the polar side of these foci and eastward on the equatorial side. As for the additional current vortex that is drawn on the night side, its longitudinal extent is very much exaggerated by the present choice of the baseline level. Since in middle and low latitudes the ionospheric conductivity in the nighttime is more than an order of magnitude lower than the daytime conductivity, the nighttime vortex would in fact be much weaker, and the daytime vortex would occupy a much greater area than is depicted in Figure 95 (Matsushita, 1965).

The equivalent current system of Sq is compared in Figure 96a with the electric field observations (solid arrows) by the Barium-cloud method. If the equivalent current indeed represents the ionospheric current driven by the ionospheric electric field that exists in the frame of reference rotating with the earth, the equivalent current should have a component (namely the Pedersen current) in the direction of the electric field. The present comparison reveals, however, that this is not the case; the equivalent current tends to have a component directed opposite to the electric field. Although the Sq-current system referred to here is the one in which the daily mean value is taken as the baseline level, it is obvious that the

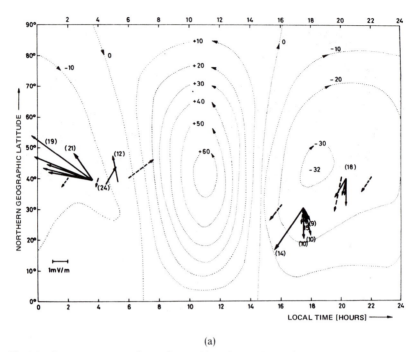

(a)

Fig. 96a. Sq current pattern for equinoxes (dotted curves) and theoretical (dashed arrows) and measured (solid arrows) electric fields (Haerendel and Lüst, 1968)

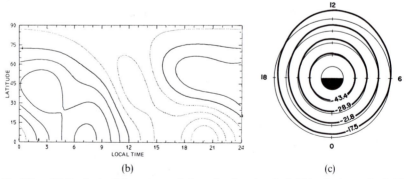

(b)   (c)

Fig. 96b,c. (b) Equipotential contours of the quiet-day electric field in the ionospheric F region. Potential maximum and minimum, which occur at the equator, are 3.3 kV and −0.5 kV, respectively. (Values above 65° latitude have no significance). (c) Contours of the total electrostatic potential in the magnetospheric equatorial plane in the unit of kV (Richmond, 1976)

foregoing discrepancy becomes even more emphasized when the day-side Sq current vortex is enlarged to occupy the entire day. It follows therefore that the ionospheric current driven by the electric field, such as the field applied from the magnetosphere, cannot be the entire source of the Sq variation (Haerendel and Lüst, 1968). The same conclusion has been reached also in the auroral zone from the analysis of the Ba-cloud observations during quiet intervals, while during disturbances the same analysis has clearly indicated that electric field of magnetospheric origin is present in the ionosphere (Wagner, 1971).

Nevertheless, rocket observations of the magnetic field have yielded abundant evidence to justify the belief that the current responsible for the Sq variation flows mainly in the ionosphere (e.g., Yabuzaki and Ogawa, 1974). Moreover, the local time dependence of the surface field produced by the magnetopause, plasma-sheet, and ring currents has been estimated to be too small compared with the observed Sq amplitude (Olson, 1970), and indeed the field from these sources seems to be too variable to constitute a regular variation like Sq. In order to explain the Sq variation, therefore, an ionospheric electromotive force has to be found that is undetectable to an observer positioned in a frame of reference fixed to the earth. Such an electromotive force was in fact envisaged a long time ago by Stewart, the essential ingredient being the wind motion in the ionosphere. Since the electric conductivity that relates the electric current to the electric field is defined in the frame of reference in which the neutral constituent is at rest, the Ohmic law that should be used in the ionosphere when the neutrals are moving with a mean velocity u is $J = (\Sigma)(E + u \times B)$. This means practically that an electromotive force $u \times B$ is produced in the ionosphere due to the presence of the neutral wind, and the effect has been known as the 'ionospheric wind dynamo' because the wind motion across the magnetic field acts as a generator. In this case E, the electric field observable in the frame of reference corotating with the earth, is the secondary product of the polarization due to the nonuniformity of the ionospheric conductivity. Namely, E is produced to satisfy div $([\Sigma](E + u \times B)) = 0$ after u is specified, and the current may well flow opposite to E. The dashed arrows in Figure 95a represent the E vectors derived by Maeda (1955) from the Sq current system on the basis of the dynamo theory, and good agreement is noted on the evening side.

On quiet days when disturbances of the magnetospheric origin are weak or absent, the foregoing electric field of ionospheric dynamo origin has been observed to stand out to distances of $L \sim 6$ or more. Figure 96b shows the equipotential contours of the quiet-day electric field in the F region. The Figure is constructed by summarizing observations of the $E \times B$ drift motion made by the whistler technique and by the backscatter method, and the contours on the positive or negative side of the arbitrarily defined zero potential are depicted by solid or dotted curves. The contour interval is 1 kV. Figure 96c contains the corresponding equipotential contours at the equatorial plane; the electric field associated with the earth's corotational motion has been added to the projection of the Sq electric field. It is seen that the equatorial drift motion during quiet intervals is directed slightly outward in the midnight-to-noon hemisphere and inward in the noon-to-midnight hemisphere (Richmond, 1976).

Much effort has been made to clarify the nature of the ionospheric wind that specifies u, and the importance of the tidal motion in the lowest-order solar diurnal mode, called $(1, -1)$ or $(1, -2)$ mode, has been recognized. The tide of this mode is driven by the in situ solar heating in the upper E and F_1 ionospheric regions, and it does not show much variability from day to day. Tidal motions of other modes, such as the $(2, 4)$ mode, are also excited, and their variability seems to be responsible for the day-to-day variability of the observed Sq field (Richmond et al., 1976).

The magnitude of the Sq field is sharply enhanced following large solar flares due to the enhancement in the ionospheric conductivity produced by the UV radiation emitted from the flare. This phenomenon is known as sfe (solar flare effect) (Van Sabben, 1968).

In addition to Sq, there is another type of geomagnetic daily variation, L, which is due to the dynamo effect of the lunar tide. The amplitude of the L variation, however, is small; it is only about 3% of that of Sq (Matsushita and Maeda, 1965). The daily variation observed during disturbances minus Sq is called S_D, while the snapshot of the local-time dependent part of the disturbance field is called DS. S_D and DS are generally not identical because the disturbance field changes with time as the disturbance develops and decays. S_D and DS are considered to comprise DP1, DP2, and the asymmetric part of the ring current field, but there might be other constituents whose natures have not yet been identified.

Equatorial Enhancement

The amplitude of Sq shows a strong enhancement at the dip equator; in the longitude sector of south America, the Sq amplitude is about two times higher at the dip equator than at the dip latitude of $4°$ (Rastogi, 1962). This enhancement has been attributed to the polarization produced within the equatorial ionosphere in the daytime. Since the conductive region is limited to a thin layer, the vertical Hall current at the dip equator (where the magnetic field is horizontal) produces polarization charges at lower and upper boundaries of the conductive region, and the charge accumulation continues until the vertical current becomes suppressed by the resulting polarization field. Thus at the equator the vertical electric field E_z is produced from the eastward electric field E_y' to satisfy

$$0 \approx \sigma_P E_z - \sigma_H E_y' \tag{108}$$

where E' is the electric field in the frame of reference moving with the neutral constituents. As a consequence, the eastward component of the electric current is augmented:

$$j_y = \sigma_P E_y' + \sigma_H E_z$$

$$= \left(\sigma_P + \frac{\sigma_H^2}{\sigma_P} \right) E_y' \tag{109}$$

Thus the effective conductivity at the dip equator is given by the Cowling conductivity (Baker and Martyn, 1953). This effect has been confirmed by electric field observations employing the incoherent scatter radar at Jicamarca (Balsley, 1973). The equatorial enhancements of SI and DP2 are also attributable to the same mechanism.

In regions other than the dip equator the growth of the polarization is prevented by the current flowing along the field line; field lines in these regions have vertical components and the discharge can proceed swiftly due to the high electric conductivity in the direction of the magnetic field (because $\sigma_0 \gg \sigma_P, \sigma_H$). Hence the enhancement of the current cannot occur.

V. Magnetosphere as a Resonator

V.1 Introduction: Magnetic Pulsations

Over a wide range of their time scale, electromagnetic perturbations propagate in the magnetosphere as hydromagnetic waves. The lower limit of the time scale is set by the period of the Larmor gyration of the ambient protons, while its upper limit is set, essentially, by the time required for the hydromagnetic wave to propagate across the closed (namely, nontail) part of the magnetosphere. The waves of this kind have long been known from geomagnetic observations and the term magnetic pulsations has been used to designate them.

The observed range of the period of pulsations extends from fractions of a second to several minutes. Pulsations are usually classified into nine categories according to the regularity and the period. This classification, pc1 ~ pc6 and pi1 ~ pi3, is listed in column (A) of Figure 97; cases having a rather well-defined spectral peak are classified as pc (continuous pulsations), while those involving a wide spectral range are classified as pi (irregular pulsations). The central frequency of some of pcs, however, does not stay fixed and shifts systematically with time. To take account of this feature more detailed classifications have been proposed that are based on the dynamic spectrum (namely, frequency vs. time characteristic) of the events. Column (B) of Figure 97 shows such a classification proposed by Saito (1976). The need for the subclassification has been felt most acutely in the short period range for pc1 and pi1.

The distinct periodicity of the pc pulsations suggests that a resonant amplification is involved in their generation mechanism. Since the high conductivity of the earth and the ionosphere acts to reduce the electric field of the wave at ionospheric heights, hydromagnetic waves are reflected by the ionosphere. For Alfven waves (namely, hydromagnetic waves of the intermediate mode) propagating along geomagnetic lines of force, therefore, the energy of the wave is confined to a limited space between the ionospheres of both hemispheres. Hence the fundamental period of the resonant oscillation is given approximately by

$$T = 2 \int \frac{ds}{V_\mathrm{A}} \tag{110}$$

(A) MATHEMATICAL			PHYSICAL					(B) CLASSIFICATION
WAVE-FORM	PERIOD RANGE (SEC)	TYPE	TYPE	NAME	LAT. DEP. OF AMP.	DIURN. VARIA. OF AMPLI.	REF. No.	SCHEMATIC DYNAMIC SPECTRUM
CONTINUOUS	0.2~5	Pc1	PP	PEARL PULSATION	S		I	
			HMC	HYDROMAGNETIC CHORUS	A	D	2	
			CE	CONTINUOUS EMISSION	A		I	
			IPDP	INTERVAL OF PULSATIONS DIMINISHING PERIODS	S	E	I	
				OTHERS				
	5~10	Pc2	AIP	AURORAL IRREGULAR PULSATION	A	M	3	
				OTHERS				
	10~45	Pc3	Pc3	Pc3	A	M-D	I	
				OTHERS				
	45~150	Pc4	Pc4	Pc4	S	D	I	
			Pg	GIANT PULSATION	A	M	I	
				OTHERS				
	150~600	Pc5	Pc5	Pc5	A	D	I	
				OTHERS				
	600~	Pc6	TF	TAIL FLUTTERING	T	N	4	
				OTHERS				
IRREGULAR	1~40	Pi1	Spt	SHORT-PERIOD Pi	A	N	I	
			PiB	Pi BURST	A	N	I	
			PiC	Pi (CONTINUOUS)	A	M	I	
			PiD	DAYTIME Pi	A	D	5	
			Psc 1 / Psi 1	Sc(Si)-ASSOCIATED Pc1,2,3	A	D	I	
				OTHERS	A	N	I	
	40~150	Pi2	Pi2	Pi2 (FORMERLY Pt)	A	N	I	
			Psfe	Sfe-ASSOCIATED PULSATION	L	D	I	
			Psc4 / Psi4	Sc(Si)-ASSOCIATED Pc4	S	D	I	
				OTHERS				
	150~	Pi3	Psc5 / Psi5	Sc(Si)-ASSOCIATED Pc5	A	D	I	
			Psc6 / Psi6	Sc(Si)-ASSOCIATED Pc6	A	D?	4	
			Pip	POLAR IRREGULAR PULSATION	A	N	4	
			Ps6	SUBSTORM-ASSOCIATED LONG-PERIOD PULSATION	A	N	4	
				OTHERS				

Fig. 97. World of magnetic pulsations; an overview by Saito (1976)

Latitudinal dependence of amplitude.
A....maximum in the auroral zone.
S....maximum in the subauroral zone.
L....maximum in low latitudes.
T....maximum in the magnetotail.

Diurnal variation of amplitude.
M....maximum in the morning sector.
D....maximum in the daytime sector.
E....maximum in the evening sector.
N....maximum in the nighttime sector.

Reference.
1....Review by Saito (1969).
2....Kokubun (1970).
3....Review by Gendrin (1970).
4....Review by Saito (1974).
5....Morioka and Saito (1971).

where the integration is to be carried out between the conjugate iono-spheres along a given field line. Dungey (1954) and Obayashi and Jacobs (1958) showed that the period thus estimated, which depends upon the L-value of the field line concerned, falls in the period range of pc2 ~ pc5. The hydromagnetic resonance has since become a basic element of pulsation theories.

As to the mechanisms for providing energy to the waves, various possibilities have been suggested and examined. These include transmissions of waves from the solar wind, instabilities of the solar-wind magnetosphere interface, sudden compressions/expansions of the magnetosphere, and night-side magnetospheric compressions associated with the collapse of field lines of the near-tail region. Pulsations resulting from the latter two causes are observed in association with SIs and substorms described earlier.

   The growth of hydromagnetic waves can occur also at the expense of the kinetic energy of charged particles residing inside the magnetosphere. Waves interact strongly with those particles with which the Doppler-shifted frequency of the wave is observed at some characteristic frequency. In the case of the cyclotron resonance this condition is expressed as

$$\omega - \boldsymbol{k} \cdot \boldsymbol{u} = \pm \Omega \qquad (111)$$

where $\boldsymbol{u}$ is the velocity of the particle and Ω is the gyrofrequency for the particle concerned. The interaction can lead to the growth of waves when the pitch-angle distribution of the particles is sufficiently anisotropic. Since the angular frequency ω and the wave vector $\boldsymbol{k}$ are related by the dispersion relation, Eq. (111) specifies the spectrum of the excited wave once the velocity distribution function is asigned. The idea of the cyclotron resonance was developed first to explain the wave emissions in the VLF range but was applied to magnetic pulsations soon afterward.

   This Chapter is concerned with the physical principles that underlie the magnetic pulsation phenomena and thus explores the basis of the diagnostic usages of magnetic pulsations. The modern theory of the hydromagnetic resonance is reviewed first, and the mechanisms that provide energy to the resonant waves are examined. Wave–particle interaction in the hydromagnetic wave range is discussed next, and finally the knowledge required to relate the magnetic-field observations on the ground to hydromagnetic wave characteristics above the ionosphere is reviewed. It is not intended to present a comprehensive description of the morphology of magnetic pulsations; useful reviews on that subject can be found in Saito (1969), Jacobs (1970), and Orr (1973).

V.2 Hydromagnetic Resonance of the Magnetosphere

The eigen period of the field-line oscillations given by Eq. (110) varies from one field line to another; it is dependent not only on the length of the field line but also on the field strength and the plasma density along the field line concerned. If all these field lines can exert resonant oscillations independently of each other and hence each field line can oscillate with its own eigen period, the dominant period of magnetic pulsations would be observed to vary continuously from place to place. Figure 98 illustrates how the period is expected to vary with latitude if the preceding is the case.

   In fact, however, theoretical examinations have demonstrated that oscillations of neighboring field lines cannot occur independently except in some special circumstances. Thus, in the magnetosphere, field lines that

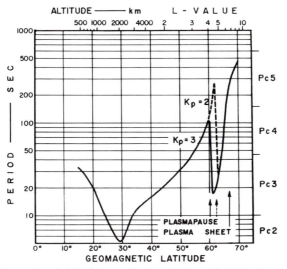

Fig. 98. Fundamental period T of the torsional oscillation of individual geomagnetic field lines. The magnetic field is represented by the dipole and field lines are labeled by the geomagnetic latitude of their intersection with the earth (see bottom) or by their equatorial crossing distance (see top). Period ranges of pc waves are indicated to the right of the figure. (Saito, 1976)

have different harmonic tones are coupled together, and the eigen period T of each field line is not necessarily reflected in the observation. Indeed the observations of pc waves have demonstrated that pulsations having the same periodicity can be registered simultaneously over latitude ranges of several degrees or more, indicating that field lines oscillate as a mass (Hirasawa, 1970; Samson and Rostoker, 1972). Hence the period T, given by Eq. (110), does not immediately explain the periodicity of the pc waves, and mechanisms have to be sought that make certain spectral peaks stand out in the pc spectrum. Such mechanisms can be classified into two types according to whether the mechanism operates inside or outside the magnetosphere. If the mechanism is external, it suffices to study the response of the magnetosphere to a monochromatic source wave. In that case Alfven waves are excited in the region where T coincides with the period of the source wave, as is shown below (McClay, 1973; Southwood, 1974; Hasegawa and Chen, 1974).

Resonant Excitation of Alfven Waves

For the sake of simplicity all the field lines are assumed to be straight and have the same length in the unperturbed state. The field direction is taken as the z axis. The medium is assumed to be inhomogeneous in the x direction, which is considered to represent the radially outward direction in the actual magnetosphere, and the y axis corresponds to the eastward direction. Background magnetic field and plasma density are represented by B and ρ, and perturbations in electromagnetic fields are expressed as

$$E = \left(E_x(x), E_y(x), 0\right)e^{i(\lambda y + kz - \omega t)}$$

$$b = \left(b_x(x), b_y(x), b_z(x)\right)e^{i(\lambda y + kz - \omega t)}$$

where the angular frequency ω and the wave vector, whose components are λ and k, are thought to be specified by the source wave. Since the length l of field lines is finite, the wavelength $2\pi/k$ parallel to the magnetic field has to be equal to, or an integral fraction of, $2l$. The electric current perpendicular to the magnetic field is given by Eq. (72) and (77), namely

$$j_\perp = \frac{1}{B^2}\left[B \times \nabla p\right] + \frac{\rho}{B^2}\frac{\partial E_\perp}{\partial t},$$

when terms that depend nonlinearly on E are neglected. If we assume further that the plasma is cold and neglect the force due to the pressure gradient in comparison with the Lorentz force and the inertia force, $j_\perp$ is given simply by

$$j_\perp = -\left(E_x(x), E_y(x)\right)\frac{i\omega\rho}{B^2}e^{i(\lambda y + kz - \omega t)}.$$

Substitution of the preceding expressions for E, b and j into rot $b = \mu_0 j$ yields

$$-\frac{\mu_0 i\omega\rho}{B^2}E_x = i\lambda b_z - ikb_y \tag{112}$$

$$-\frac{\mu_0 i\omega\rho}{B^2}E_y = ikb_x - \frac{db_z}{dx}. \tag{113}$$

Using $-\partial b/\partial t = $ rot E to relate E and b, we obtain

$$\left(\frac{\mu_0\rho}{B^2}\omega^2 - k^2\right)E_x = i\lambda\left(\frac{dE_y}{dx} - i\lambda E_x\right) \tag{114}$$

$$\left(\frac{\mu_0\rho}{B^2}\omega^2 - k^2\right)E_y = -\frac{d}{dx}\left(\frac{dE_y}{dx} - i\lambda E_x\right). \tag{115}$$

The equation to be satisfied by E_y is, therefore,

$$\frac{d}{dx}\left(\frac{K^2 - k^2}{K^2 - k^2 - \lambda^2}\frac{dE_y}{dx}\right) + (K^2 - k^2)E_y = 0 \qquad (116)$$

where $K^2 \equiv (\mu_0\rho/B^2)\omega^2 \equiv \omega^2/v_A^2$.

If account had been taken of the pressure effect by the adiabatic approximation (with the adiabatic exponent represented by γ), the equation for E_y would have been

$$\frac{d}{dx}\left(\frac{\alpha B^2(K^2 - k^2)}{K^2 - k^2 - \alpha\lambda^2}\frac{dE_y}{dx}\right) + B^2(K^2 - k^2)E_y = 0 \qquad (117)$$

where

$$\alpha = 1 + \gamma\beta + \frac{\gamma^2\beta^2 k^2}{K^2 - \beta\gamma k^2}$$

and $\beta = \mu_0 p/B^2$ (Hasegawa and Chen, 1974). Although this equation is more rigorous, we shall nevertheless adopt Eq. (116) for further examination since around the singularity at $K^2 - k^2 = 0$ (with which we shall be concerned), the preceding two equations give essentially the same result.

Eq. (116) can be transformed to

$$\frac{d^2 E_y}{dx^2} - \lambda^2\frac{dK^2}{dx}\frac{1}{(K^2 - k^2)(K^2 - k^2 - \lambda^2)}\frac{dE_y}{dx} + (K^2 - k^2 - \lambda^2)E_y = 0. \qquad (118)$$

If the medium is homogeneous and hence $dK^2/dx = \mu_0\omega^2 d(\rho/B^2)/dx = 0$, or if the wave vector is confined to the xz plane and hence $\lambda = 0$, Eq. (118) reduces to the dispersion equation for the hydromagnetic wave of the fast mode. In the presence of the inhomogeneity and the azimuthal propagation, however, the second term has a singularity at $x = x_0$ where $(K^2 - k^2)$ is zero. The condition $K^2 - k^2 = 0$ means that the resonant period $2\pi/kV_A$ of the Alfven wave propagating along the local field line matches the period $2\pi/\omega$ of the source wave, and the amplitude of E_y is made logarithmically infinite at $x = x_0$. This singularity has been interpreted to represent the excitation of the Alfven wave on the resonant field line due to the coupling with the fast-mode wave. When ω is taken to have a positive imaginary part ω_i, $K^2 - k^2$ can be expanded around $x = x_0$ as

$$K^2 - k^2 = (x - x_0)\left.\frac{dK^2}{dx}\right|_{x=x_0} + i\varepsilon\left.\frac{dK^2}{dx}\right|_{x=x_0} \qquad (119)$$

where

$$\varepsilon = -\frac{\omega_i}{\omega_r} \frac{1}{d(\ln V_A)/dx}\bigg|_{x=x_0}.$$

Thus around $x = x_0$ Eq. (118) becomes

$$\frac{d^2E_y}{dx^2} + \frac{1}{x - x_0 + i\varepsilon} \frac{dE_y}{dx} - \lambda^2 E_y = 0 \qquad (120)$$

and its solution is given by

$$E_y(x) = CI_0(\lambda(x - x_0 + i\varepsilon)) + DK_0(\lambda(x - x_0 + i\varepsilon)) \qquad (121)$$

which has a maximum but stays finite at $x = x_0$. I_0 and K_0 are modified Bessel functions. The solution is shown schematically in Figure 99; the fast-mode wave with a monochromatic spectrum (which is specified by the external source) transports energy through the magnetosphere, because the wave of this mode can propagate across the magnetic field, and produces the large-amplitude Alfven wave in the region where the resonance condition is met. In some models the fast-mode wave concerned is assumed to have such a short wavelength in the y direction ($K^2 \ll \lambda^2$) that it has a character of the surface wave. The decay of the wave amplitude illustrated in the neighborhood of the magnetopause in Figure 99 is due to this assumption. [The transport of energy by an evanescent wave (such

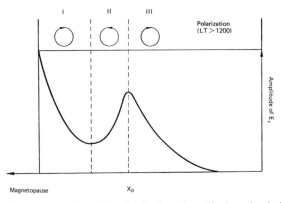

Fig. 99. Schematic representation of the distribution of amplitude and polarization of the hydromagnetic wave in the model magnetosphere. A monochromatic surface wave excited on the magnetopause is supposed as the source and x_0 denotes the resonant point (Southwood, 1974)

as the surface wave) to the resonant point is described in Stix (1962) as 'Budden tunneling.']

The polarization of the wave is given by

$$\frac{b_y}{b_x} = -\frac{E_x}{E_y} = -\frac{i\lambda}{(K^2 - k^2 - \lambda^2)} \frac{dE_y}{dx} \frac{1}{E_y}. \tag{122}$$

Because the time dependence is expressed as $\exp(-i\omega t)$, the polarization with respect to the z direction is right-handed (left-handed) when the imaginary part of (b_y/b_x) is positive (negative). The occurrence of the resonance affects the spatial distribution of the polarization as $d(\ln E_y)/dx$ changes sign at the resonant point. In the region where the fast-mode wave has the character of the surface wave, $K^2 - k^2 - \lambda^2 < 0$. Hence if the sense of the polarization is described for one looking along the background field, it is clockwise or counterclockwise, according to whether $\lambda d(\ln E_y)/dx$ is positive or negative. This polarization in the northern hemisphere is shown schematically in Figure 99 for the case of an eastward propagating wave ($\lambda > 0$). In the neighborhood of the magnetopause (region I) where the amplitude of the surface wave falls off, $d(\ln E_y)/dx > 0$ and the polarization is clockwise. The polarization becomes counterclockwise as the resonant point of the Alfven wave is approached (in region II), and it is linear at the resonant point where the amplitude is maximum. Then it turns clockwise again earthward of the resonant point (in region III). For the westward propagating surface wave the polarization is expected to be opposite to those indicated in Figure 99.

It would be worthwhile to note here that Eq. (122) was derived only from Eq. (114) and Faraday's law. The applicability of the foregoing polarization law is therefore not limited to a specific resonance process such as the fast-Alfven coupling.

Long-Period Magnetic Pulsations

Figure 100 shows local-time and latitude dependences of the polarization characteristics of pc5 waves. The observation was made by the Canadian network of stations, where LGT (Geomagnetic local time) is UT minus 0830. At all the stations belonging to this chain, the reversal of the polarization is observed twice a day, namely around 1130–1230 LGT (2000–2100 UT) and around 1830–1930 LGT (0300–0400 UT). Below the latitude of the amplitude maximum (which is indicated by a dashed curve), the polarization is counterclockwise (indicated by a solid ellipse) in the midnight-to-morning sector and is clockwise (indicated by a dotted ellipse) in

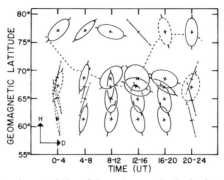

Fig. 100. Polarization characteristics of the pc5 waves in the horizontal plane. Centers of the ellipses are at the latitudes of the stations from which the data were obtained. Local time of the stations is roughly UT minus 8.5h. Solid and dotted ellipses correspond to counter-clockwise and clockwise polarizations. Sizes of ellipses are not related to intensity. (Samson, 1972)

the afternoon sector. Since these latitudes correspond to region III of Figure 99, the comparison with Eq. (122) indicates that waves are propagating westward ($\lambda < 0$) in the former sector and eastward ($\lambda > 0$) in the latter; on the day side this corresponds to the wave propagation away from the nose of the magnetosphere. It is noteworthy that the diurnal variation in the latitude of the maximum pc5 amplitude runs parallel to the auroral oval (Samson, 1972).

Simultaneous observations of a pc5 event, made on the ground at a conjugate pair of high-latitude stations and by a satellite located in the neighborhood of field lines that pass these ground stations, are displayed in Figure 101. The observations were made in the morning sector around 0600 LT when the satellite was located at geomagnetic latitudes of $13° \sim 9°$. Let us compare the behavior of the pc5 waves that were clearly recorded (with periods of $200 \sim 300$ s) at every site. First, the comparison of waves at the conjugate ground stations shows that variations in the H component are in phase but those in the D component are out of phase. This means that the motions of the field line at the northern and the southern ends are symmetric with respect to the equatorial plane. Second, the amplitude of the waves is $3 \sim 6$ times higher on the ground than at the equatorial satellite position, and the survey of the transverse pc5 waves by OGO 5 has revealed that waves are detected mainly at magnetic latitudes above $10°$, indicating that the equatorial plane is the nodal plane for the transverse component of the pc5 waves (Kokubun et al., 1976). These two points strongly suggest that pc5 waves represent odd mode standing waves. The odd mode concerned is likely to be the fundamental mode, since the observed period is very close to the fundamental period T

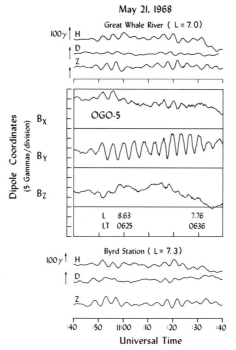

Fig. 101. Simultaneous observations of pc5 waves at a conjugate pair of auroral-zone stations (top and bottom) and at a satellite located on a nearby field line (middle) (Kokubun et al., 1976)

estimated for high latitudes. (For the fundamental mode the field-line length is one half of the wavelength.) The standing nature of the pc5 waves has been confirmed by the 90° or 270° phase difference that has been detected by OGO 5 between magnetic field perturbations and thermal-particle flux variations; such phase relation means that the time-averaged Poynting flux along the ambient magnetic field is zero (Kokubun et al., 1977).

The intensity maximum of pc5 waves tends to be found at a lower latitude when the Kp index is higher, and the dominant wave period varies from ~ 500 s to less than 300 s as the intensity maximum is shifted from the geomagnetic latitude of $\sim 70°$ to $\sim 60°$ (Hirasawa, 1970; Samson and Rostoker, 1972). This relation between the period and the latitude is consistent with the T vs. latitude relation depicted in Figure 98, and it indicates that the pc5 resonance occurs on an increasingly inner field line when the magnetosphere is more agitated. In almost all cases of *individual* pc5 activity, however, pc5 waves are observed to have the same period wherever it is observed; as discussed at the beginning of this Section, it is not the individual field lines but the collection of them that exerts the resonant oscillation.

pc3 and 4 waves also show high frequencies of occurrence at latitudes where the resonant period T shown in Figure 98 falls in the respective period ranges. Figure 102 shows the occurrence frequency of dominant pc3 and 4 periods at various latitudes (Hirasawa, 1969). It is seen that pc3 pulsations with periods of about 30 s tend to occur at two separate latitude ranges, namely, above $\sim 58°$ and below $\sim 50°$; in Figure 98 these latitude ranges correspond to the plasmatrough (i.e., the region outside the plasma-pause) and the low-latitude region where waves with pc3 periods are indeed expected to resonate. pc4 waves with periods of about 60 s, on the other hand, tend to be observed in the middle latitude range sandwiched between the two pc3-dominating regions; since $\sim 60°$ is the latitude where the plasmapause is formed under normal conditions, this is indeed the latitude range where the period T is expected to be in the pc4 range. Closer investigations have been made of the latitude where the dominant wave type changes from (high-latitude) pc3 to pc4, and this has been found to show local-time and Kp dependences similar to those of the plasmapause

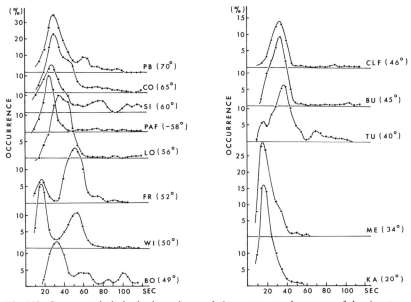

Fig. 102. Geomagnetic latitude dependence of the occurrence frequency of dominant pc periods. The data used are those obtained during the IGY (Hirasawa, 1969)

(Orr and Webb, 1975). The latitudinal variations in intensity and polarization of pc3 and 4 waves also support the view that the resonant regions of (high-latitude) pc3 waves and of the pc4 waves are located within the plasmatrough and inside (or, at) the plasmapause, respectively (Lanzerotti et al., 1974). However, the occurrence in middle latitudes (50° ~ 55°) of pc3 waves with a period of ~ 15 s does not have a corresponding feature in Figure 98.

The polarization ellipses of both high-latitude pc3 waves and pc4 waves at a conjugate pair of stations have been found to be mirror images with each other with respect to the meridional plane, as would be the case when the motion of field lines is symmetrical with respect to the equatorial plane. Hence the observation supports the view that these waves are also odd mode (probably fundamental mode) standing waves. Below the latitudes of the respective intensity maxima, the sense of polarization of magnetic pulsations having periods of ~ 40 to ~ 70 s tends to be clockwise from ~ 1200 LT to ~ 1800 LT and counterclockwise at other local times (as viewed along the background field). On the day side this indicates the wave propagation away from the noon meridian and on the night side from the local morning toward the local evening across midnight (Lanzerotti et al., 1976).

Surface Wave at V_A Discontinuity

Let us return to the theoretical discussion of the mechanism that gives well-defined periodicity to pc waves and examine, this time, processes internal to the magnetosphere. A possibility to be considered is the excitation of the surface wave on a surface where the Alfven velocity changes abruptly. Since such a wave has an eigen period of its own, magnetic pulsations with a discrete spectral peak would be generated when the surface is subjected to broad-band oscillations whose spectrum covers that peak. To estimate the eigen period of this surface wave, let us assume that ρ and B change discontinuously across a plane located at $x = x'$ and discriminate the quantities on each side by using suffixes 1 and 2. Since both ρ and B are assumed to be uniform on each side of the discontinuity, the equation for E_y is reduced to Eq. (118) without the second term:

$$\mu_i^2 + K_i^2 - k^2 - \lambda^2 = 0, \qquad i = 1, 2 \tag{123}$$

where x dependence of the perturbation is expressed as $e^{-\mu_i|x-x'|}$, where $\mu_i > 0$. So that the plasmas on both sides of the discontinuity do not separate, their velocities $u_x = E_y/B$ perpendicular to the discontinuity should be identical, namely,

$$\frac{E_{1,y}}{B_1} = \frac{E_{2,y}}{B_2} \tag{124}$$

and it is also required that waves on both sides have the same values of k and λ. The pressure balance condition across the discontinuity is given by

$$B_1 b_{1,z} = B_2 b_{2,z} \tag{125}$$

since the plasma is assumed to be cold. By using Eq. (113) and $i\omega b_x = -ik E_y$, b_z can be replaced by E_y;

$$b_z = \mp \frac{i\omega}{\mu_i} \left(\frac{1}{V_A^2} - \frac{k^2}{\omega^2} \right) E_y \tag{126}$$

where $-$ or $+$ sign is adopted according to $x > x'$ or $x < x'$. To satisfy Eqs. (124) and (125), therefore

$$\frac{B_1^2}{\mu_1} \left(\frac{1}{V_{A,1}^2} - \frac{k^2}{\omega^2} \right) = -\frac{B_2^2}{\mu_2} \left(\frac{1}{V_{A,2}^2} - \frac{k^2}{\omega^2} \right). \tag{127}$$

Now if the azimuthal wavelength is sufficiently small so that $K_i^2 = \frac{\omega^2}{V_{A,i}^2} \ll \lambda^2$ for both $i = 1$ and 2 sides, we can approximate Eq. (123) by $\mu_1 = \mu_2 = |\lambda|$. It follows then that

$$\omega = \left(\frac{B_1^2 + B_2^2}{\mu_0 (\rho_1 + \rho_2)} \right)^{1/2} k. \tag{128}$$

Since k is supposed to be fixed by the length of the field line, Eq. (128) gives the eigen frequency of the surface wave on the discontinuity. When the change in V_A is due to the change in density, as is the case at the plasmapause, the eigen period T_s of the surface wave is

$$T_s = \frac{2l}{\sqrt{2} V_{A,2}} \tag{129}$$

where the suffix 2 designates the value on the high density side of the discontinuity and l is the length of the field line.

A more detailed examination of this resonance, in which the finite thickness of the V_A discontinuity is taken into account, yields the imaginary part of ω:

$$\frac{\omega_i}{\omega_r} \simeq -\frac{\pi}{2}(\lambda \cdot 2a)\frac{\mu_0 \rho_1 \rho_2 (V_{A,1}^2 - V_{A,2}^2)}{(\rho_1 + \rho_2)(B_1^2 + B_2^2)} \tag{130}$$

where $2a \simeq 1 \left/ \left|\dfrac{d(\ln \rho)}{dx}\right|_{x=x'}\right.$ represents the scale length of the discontinuity.
ω_i is made negative, namely, the wave damps, due to the phase mixing effect (Hasegawa and Chen, 1974). At a discontinuity of ρ but not of B, Eq. (130) gives $\omega_i/\omega_r \simeq -\frac{\pi}{4}(\lambda \cdot 2a)(\rho_2 - \rho_1)/(\rho_2 + \rho_1)$, and hence the excited wave lasts longer when the density discontinuity is smaller and also when the thickness of the discontinuity is smaller. Thus the wave excitation on the density discontinuity does not require the density jump to be large, and hence surface waves would be generated not only on the plasmapause but also on other sharp but minor density nonuniformities that may be prevalent.

Since the eigen periods defined by T and T_s do not differ very much, observed pc waves can in general be attributed equally well to a resonance excited by a monochromatic external source or to an excitation of a surface wave at a sharp density gradient by a broad-band noise like a pulse. In order to distinguish these two possibilities it is necessary to have a precise determination of the position of the intensity maximum and/or the polarization reversal relative to the density discontinuity. Such an attempt has been made for waves having periods of 100 to 30 s observed in pre-midnight hours (2100 ~ 2400 LT) (Lanzerotti and Fukunishi, 1975). Figure 103 is a comparison of the latitudinal profile of the electron density, measured by ISIS 2 at ~ 1400 km altitude, with wave polarizations observed on the ground in nearly the same meridian. The location of the equatorial plasmapause in the late local evening sector is also indicated by the solid bars; the Explorer 45 electric field observations are used for its determination. In cases (a) and (b) the reversal in the wave polarization with latitude may be associated with electron density decreases that began at $L \sim 4.2$ and 3.6, respectively, but in case (c) no particularly sharp density gradients were observed at the polarization reversal. Since it is known that the latitude profile of the topside electron density does not precisely reflect the plasma distribution at higher altitudes, it is important to repeat these kinds of analyses when light ion data become available in higher altitudes on field lines passing through chains of stations observing magnetic pulsations.

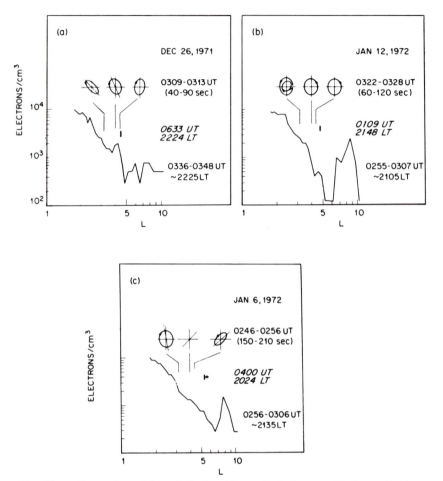

Fig. 103a–c. Comparison of the polarization ellipses of pc3–4 waves with the topside electron density distribution in the nearby meridian (Lanzerotti and Fukunishi, 1975)

The preceding analysis of Lanzerotti and Fukunishi (1975) has disclosed, in addition, that the location of the intensity maximum and that of the polarization reversal do not always agree for the nighttime pulsation events. This may be due to the presence of multiple resonant peaks within the spectral range studied, or to the effect of the azimuthal pressure gradient that is neglected in Eq. (122). The possibility of the ionospheric modification (see Section V.5) would also have to be considered. More information on the field and particle distributions in the region of the wave excitation is apparently needed for the improvement of the resonance model.

Another possibility of the internal mechanism to produce spectral peaks of pc waves is the trapping of the fast-mode wave between high values of the phase velocity of the fast-mode wave (Tamao, 1969). As noted in Figure 9, V_A (which approximates the fast-mode velocity) has a maximum in the plasmatrough and an increasing trend toward the earth. For the fast-mode wave propagating perpendicular to the magnetic field the reflection occurs at the point $x = x''$ where $\omega^2/V_A^2(x'') - \lambda^2 = 0$. In a one-dimensional model the resonant period is given by $T = 2 \int_{x_1'}^{x_2'} \dfrac{dx}{V_A}$, where x_1'' and x_2'' are two values, one in the plasmatrough and the other inside the plasmasphere, of x''. The resonant wave produced in this case is standing wave in the radial direction.

V.3 Energy Source of Hydromagnetic Waves

Like other magnetospheric phenomena, magnetic pulsations are thought to be driven by the energy supplied from the solar wind. The energy transferred to the magnetosphere by the day-side reconnection process would obviously be one of the major causes of the wave generation, but, in addition, hydromagnetic interactions at the magnetopause seem to play an important role in producing magnetic pulsations.

The hydromagnetic interactions that can give rise to hydromagnetic waves inside the magnetosphere are the transmission of the waves from the magnetosheath and the excitation of the waves at the magnetopause. Satellite traversals through the magnetosheath have demonstrated that the magnetosheath is a highly wavy region. These magnetosheath waves, which are probably associated with the dissipation process operating at the bow shock, can act as a powerful source of magnetic pulsations if they can be transmitted through the magnetopause with enough efficiency. The presence of the solar-wind flow on the magnetosheath side, on the other hand, means that the magnetopause is a boundary layer with a velocity shear. Waves would hence be excited on the magnetopause by the Kelvin-Helmholtz instability if the local velocity of the solar wind just outside the magnetopause exceeds a given threshold.

Wave Transmission from the Magnetosheath

Let us represent the undisturbed magnetopause by a plane surface at $x = 0$ and express the undisturbed flow velocity V and magnetic field B as

$$V = (0, V_y, V_z), \qquad B = (0, B_y, B_z).$$

The solar wind occupies the region $x < 0$ and the magnetosphere is on the $x > 0$ side; the variables belonging to these two regions will be distinguished by subscripts 1 and 2. Perturbations are assumed to have space and time dependences of the plane-wave type: $\exp(-i(\omega t - k_x x - k_y y - k_z z))$ and variables associated with incident, reflected, and transmitted waves are denoted by superscripts (i), (r), and (t), respectively. The perturbation x_I of the x coordinate of the interface is written as

$$x_I = a e^{-i(\omega t - k_t \cdot r)}$$

where a is the amplitude of the oscillation and $k_t = (0, k_y, k_z)$.

The pressure balance across the interface requires that

$$p^{(i)} + p^{(r)} + \frac{B_1}{\mu_0} \cdot (b^{(i)} + b^{(r)}) = p^{(t)} + \frac{B_2}{\mu_0} \cdot b^{(t)} \tag{131}$$

where, unlike the previous Eq. (125), the thermal pressure is included, because this is important on the magnetosheath side. For each component of the wave, there is a relation

$$p + \frac{B \cdot b}{\mu_0} = \rho \frac{\omega'^2 - (k_t \cdot V_A)^2}{k_x \omega'} v_x \tag{132}$$

where $E_t = (0, E_y, E_z)$ and $\omega' = \omega - k \cdot V$ is the wave frequency observed in the rest frame of the fluid. The preceding equation follows Eq. (113) when the current due to the pressure gradient is included. Hence Eq. (131) can be rewritten as

$$\rho_1 \frac{(\omega_1'^2 - (k_t \cdot V_{A,1})^2)}{\omega_1'} \left(\frac{v_x^{(i)}}{k_x^{(i)}} + \frac{v_x^{(r)}}{k_x^{(r)}} \right) = \rho_2 \frac{(\omega_2'^2 - (k_t \cdot V_{A,2})^2)}{\omega_2'} \frac{v_x^{(t)}}{k_x^{(t)}}. \tag{133}$$

The condition that the flow velocity normal to the interface is zero is expressed as

$$v_x^{(i)} + v_x^{(r)} - V_1 \cdot \text{grad } x_I - \frac{\partial x_I}{\partial t} = 0$$

$$v_x^{(t)} - V_2 \cdot \text{grad } x_I - \frac{\partial x_I}{\partial t} = 0. \tag{134}$$

We can rewrite the foregoing as

$$\frac{v_x^{(i)} + v_x^{(r)}}{\omega_1'} = \frac{v_x^{(t)}}{\omega_2'}. \tag{135}$$

The transmission coefficients are therefore

$$T(v) \equiv \frac{v_x^{(t)}}{v_x^{(i)}} = \frac{\omega_2'}{\omega_1'} \frac{2}{1 + Z}$$

$$T(b) \equiv \frac{b_x^{(t)}}{b_x^{(i)}} = \frac{(k_t \cdot B_2)}{(k_t \cdot B_1)} \frac{2}{1 + Z}$$

where

$$Z \equiv \frac{k_x^{(i)} \rho_2 (\omega_2'^2 - (k_t \cdot V_{A,2})^2)}{k_x^{(t)} \rho_1 (\omega_1'^2 - (k_t \cdot V_{A,1})^2)} \tag{136}$$

and the relation $k_x^{(r)} = -k_x^{(i)}$ has been used (McKenzie, 1970).

k_x in the foregoing expression of the transmission coefficient is related to k_t and ω' by the dispersion equation of hydromagnetic waves. Among the three modes of the hydromagnetic waves, the Alfven (intermediate) mode can be excluded from the present consideration since this mode carries energy along the magnetic field only; even if the component B_x of the magnetic field normal to the interface is nonzero in reality due to the reconnection effect, the Alfven wave would be transmitted to the polar cap only. The dispersion equation for the remaining two modes is known to be

$$k_x^2 = -k_t^2 + \omega'^4 \{\omega'^2 (V_s^2 + V_A^2) - V_s^2 (k_t \cdot V_A)^2\}^{-1} \tag{137}$$

where $V_s = (\gamma \kappa T / m)^{1/2}$ is the sound speed. In the present model the pressure balance across the unperturbed magnetopause is expressed as

$$n_1 \kappa T_1 + \frac{B_1^2}{2\mu_0} = n_2 \kappa T_2 + \frac{B_2^2}{2\mu_0}.$$

This roughly means

$$n_1 (V_{s,1}^2 + V_{A,1}^2) \sim n_2 (V_{s,2}^2 + V_{A,2}^2) \tag{138}$$

and hence $V_{s,1}^2 + V_{A,1}^2 \ll V_{s,2}^2 + V_{A,2}^2$ because $n_1 \gg n_2$ at the day-side magnetopause. Because of this inequality, Z tends to be very large. Numerical estimates of the transmission coefficients (McKenzie, 1970; Wolfe and Kaufmann, 1975) have revealed that incident waves are totally reflected unless the angle of incidence is nearly normal to the interface; if wave vectors of magnetosheath waves are distributed isotropically, only $1 \sim 2\%$ of the energy of the fast-mode wave would be transmitted into the magnetosphere. The slow-mode waves are reflected completely.

It has to be noted, however, that when the wave is totally reflected at the interface, the evanescent wave (with $k_x^{(t)2} < 0$) generally appears on the magnetosphere side. This evanescent wave, whose amplitude decays

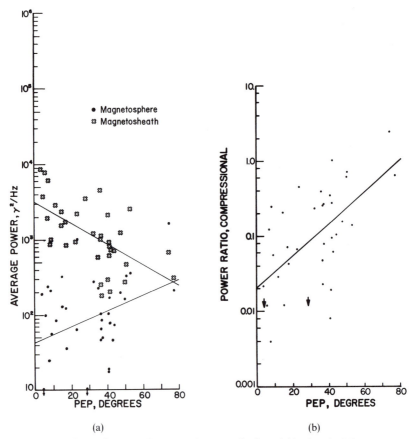

(a)   (b)

Fig. 104a,b. Observed power of compressive waves in the neighborhood of the magnetopause. (a) Power levels on the magnetospheric and the magnetosheath sides, and (b) their ratio at each magnetopause crossing (Wolfe and Kaufmann, 1975)

like exp $(-|k_x^{(t)}|x)$ away from the interface, behaves in the same way as the surface wave, and it can couple with the Alfven wave if the resonant condition is satisfied somewhere inside. Thus the energy of the magneto-sheath wave can be transmitted into the magnetosphere even when perfect reflection is indicated in a simplified model in which the region 2 is assumed to be uniform and infinite in its extent in the direction away from the interface (i.e., Budden tunneling can take place).

The power of compressive waves with periods of 0.5 to 2 min observed in the magnetosphere is compared with that detected in the magnetosheath in Figure 104. The abscissa is the angle that the satellite-earth line makes with the solar-wind direction, and data are taken from 36 magnetopause crossings of Explorer 12 in the 0600–1200 LT quadrant. Figure 104a shows the average power observed in each region and Figure 104b demonstrates their ratio. It is seen that near the sun–earth line the observed ratio is low: several percent on the average. Fast-mode waves in this region of the magnetosphere can thus be potentially attributable to the transmission of the turbulent hydromagnetic waves from the magnetosheath. Furthermore, the power spectrum of compressional waves in the magnetosheath often contains peaks, although their exact frequencies and amplitudes are variable (Fairfield, 1976). Hence there is a possibility that these magnetosheath waves are transmitted to the magnetosphere and excite monochromatic Alfven waves in the manner described in the early part of the last Section. Beyond $30° \sim 40°$ from the sun–earth line, however, the observed ratio seems to be too high to be explained by the wave transmission from the magnetosheath (Wolfe and Kaufman, 1975).

Kelvin-Helmholtz Instability

When $v_x^{(i)}$ representing the incident wave is set to zero, Eqs. (133) and (135) give

$$\frac{\rho_1(\omega_1'^2 - (\boldsymbol{k}_t \cdot \boldsymbol{V}_{A,1})^2)}{k_x^{(r)}} = \frac{\rho_2(\omega_2'^2 - (\boldsymbol{k}_t \cdot \boldsymbol{V}_{A,2})^2)}{k_x^{(t)}}. \tag{139}$$

The combination of this equation with Eq. (137) produces a tenth degree equation for ω (Southwood, 1968). When the tangential wavelength is short enough, however, Eq. (137) can be approximated by $k_x^2 + k_t^2 = 0$ that represents the surface wave. In that circumstance Eq. (139) can be simplified as

$$(\rho_1 + \rho_2)\omega^2 - 2\{\rho_1(\boldsymbol{k}_t \cdot \boldsymbol{V}_1) + \rho_2(\boldsymbol{k}_t \cdot \boldsymbol{V}_2)\}\omega + \{\rho_1(\boldsymbol{k}_t \cdot \boldsymbol{V}_1)^2$$
$$+ \rho_2(\boldsymbol{k}_t \cdot \boldsymbol{V}_2)^2 - \rho_1(\boldsymbol{k}_t \cdot \boldsymbol{V}_{A,1})^2 - \rho_2(\boldsymbol{k}_t \cdot \boldsymbol{V}_{A,2})^2\} = 0. \tag{140}$$

Since a strong flow exists only on the magnetosheath side, $V_2 = 0$, and the foregoing equation yields

$$\omega = \frac{1}{\rho_1 + \rho_2} \left[\rho_1 (k_t \cdot V_1) \pm \sqrt{(\rho_1 + \rho_2)\{\rho_1 (k_t \cdot V_{A,1})^2 + \rho_2 (k_t \cdot V_{A,2})^2\}} \right.$$
$$\left. - \rho_1 \rho_2 (k_t \cdot V_1)^2 \right]. \tag{141}$$

The surface wave becomes unstable, namely, the Kelvin-Helmholtz instability is induced, when

$$(k_t \cdot V_1)^2 > \frac{\rho_1 + \rho_2}{\rho_1 \rho_2} \{\rho_1 (k_t \cdot V_{A,1})^2 + \rho_2 (k_t \cdot V_{A,2})^2\}. \tag{142}$$

The foregoing condition is more easily met when k_t is parallel to the streaming velocity V_1 or when k_t is perpendicular to the field B_2 on the magnetospheric side. If k_t satisfies both of these optimum conditions, as it would near the equatorial plane, the inequality becomes $V_1^2 > B_1^2 \cos^2 \varphi / \mu_0 \rho_2$ where φ is the angle between k_t and B_1. $\rho_1 \gg \rho_2$ has been assumed. With $B_1 = 20\gamma$, $n_2 = \rho_2/m_p = 0.1/\text{cm}^3$ and $\varphi = 0$, the required condition becomes $V_1 > 2 \times 10^3$ km/s, which is still not easy to satisfy. For the Kelvin-Helmholtz instability to be a significant agent of the wave generation, therefore, it seems necessary that k_t be nearly perpendicular to the magnetosheath field B_1 also. It remains, however, to incorporate the effects of the day-side reconnection and the entry layer plasma on the condition of the wave excitation.

Comparison with Interplanetary Observations

The comparison of the pc3 and pc4 activities with solar-wind parameters have revealed indeed that the activities of these pulsations are controlled by the direction of the interplanetary magnetic field (Bolshakova and Troitskaya, 1968; Nourry, 1976). Figure 105 depicts the histograms showing the occurrence frequencies of pc3 and pc4 pulsations at Ralston (58°) as functions of latitude (θ_{SE}) and longitude (ϕ_{SE}) of the interplanetary magnetic field. The entire distribution of the IMF orientation in the interval studied (August to November, 1967) is taken into account in 'normalizing' the distribution. The Figure clearly shows that activities of both pc3 and pc4 waves are augmented when the IMF is directed nearly radially, i.e., when $\theta_{SE} \sim 0°$ and $\phi_{SE} \sim 0°$ or 180°, the dependence being more pronounced for pc4 than for pc3. Making a distinct contrast to the

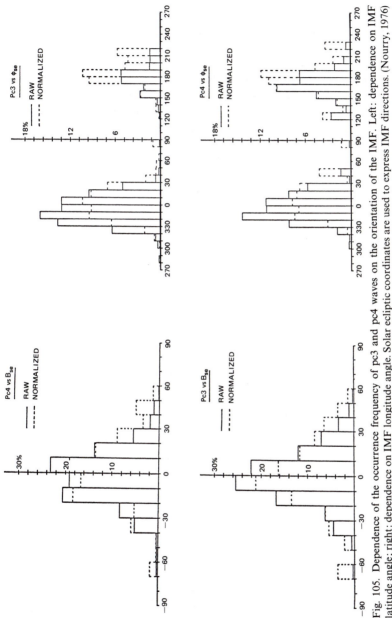

Fig. 105. Dependence of the occurrence frequency of pc3 and pc4 waves on the orientation of the IMF. Left: dependence on IMF latitude angle; right: dependence on IMF longitude angle. Solar ecliptic coordinates are used to express IMF directions. (Nourry, 1976)

features discussed in Chapters II–IV and attributed essentially to the reconnection process, the θ_{SE} dependence of Figure 105 does not show any sign of preference to the southward IMF polarity.

The foregoing observation can be compared with wave transmission and *K-H* wave excitation models discussed earlier in this Section. First, with regard to the transmission model we note that the structure of the bow shock is known to be dependent on the angle between the magnetic field and the shock normal. When this angle is small, the shock becomes a noisy wave-train boundary rather than an abrupt change (Greenstadt, 1972). Under the nearly radial IMF, therefore, waves are produced right around the subsolar region of the bow shock, and a large part of the magnetosphere makes contact with turbulent waves that are convected by the solar wind from that region. The correlation observed between the radial orientation of IMF and the activation of pc3 and 4 waves could probably be due to the transmission of these waves into the magneto-

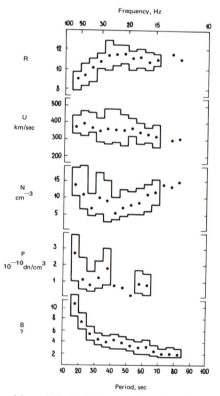

Fig. 106. Comparison of the period of pc3–4 waves at Borok and Petropavlovsk with various parameters of the solar wind (Gulelmi et al., 1973)

sphere. Second, concerning the instability model, we note that the radial IMF causes the velocity V_1 to be locally aligned with the magnetic field B_1 in the magnetosheath. This means that the direction of k_t that makes the left-hand side of the inequality (142) large makes the first term on the right-hand side large at the same time. Thus the observation does not seem to favor the Kelvin-Helmholtz instability model.

Figure 106 shows the result of another attempt to find the dependence of the pc characteristics on solar wind conditions. In this Figure the factor that influences the period of pc waves is sought, and the pc periods observed at Borok (53°) and Petropavlovsk (45°) are compared with (from the top) radius of the magnetosphere, velocity, density, pressure, and magnetic field of the solar wind (Gulelmi and Bolshakova, 1973; Gulelmi et al., 1973). It is seen that the period decreases as the strength of the interplanetary magnetic field increases, and this tendency has been interpreted to suggest that the origin of pc3, 4 waves lies in upstream waves that are excited beyond the bow shock by the bunch of reflected protons (Gulelmi, 1974). It is not clear, however, if the change in the pc period at the preceding two stations truely represents the global change in the predominant period of pc3–4 pulsations. Of particular concern is that Borok is located in the neighborhood of the plasmapause. Since pc3 is strong in the plasmatrough while pc4 is intense inside the plasmapause, the dominant type of pc at this station would be changed by the shift in the plasmapause position. It appears conceivable that a stronger IMF is correlated statistically with a smaller plasmapause radius and hence causes an increase in the probability that pc3 signals are stronger there.

Pulsations Associated with SIs and Substorms

For magnetic pulsations of some particular types, the energy source can be deduced from the geomagnetic disturbance phenomena that they accompany. For example, the sudden impulses discussed in Chapter I are known to be followed by pulsations having periods in the pc2 ~ pc5 range. The energy source of these pulsations can be clearly identified with the large-scale compressive or expansive waves that lead to compression or expansion of the entire magnetosphere. Although such a source wave has a broad spectrum, similar to a step function, magnetic pulsations induced have well-defined spectral peaks that fall in pc5, pc4, and pc3 ranges. It has been recognized that these SI-associated pulsations are very similar to pc pulsations in general; as shown in Figure 107 these similarities are noted not only in the latitude dependence of the predominant periodicities but also in the local-time dependence of amplitudes

Magnetosphere as a Resonator

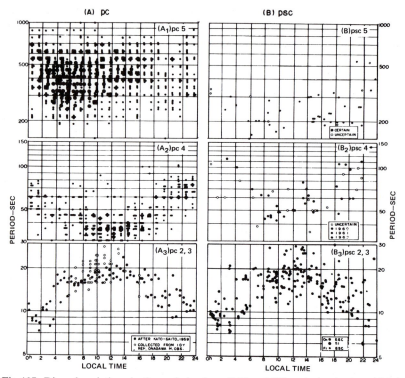

Fig. 107. Diurnal variations in the periods of pcs (left) and SC-excited pulsations (right). In each column three panels correspond, from the top, to waves in pc5, pc4 and pc3 ~ 2 period ranges. Data used are, from A_1, A_2, through B_3, observations at Byrd ($-71°$), Fredericksburg (50°), Onagawa (28°), Byrd, Fredericksburg, and Onagawa-Fredericksburg (Saito and Matsushita, 1967)

and periods (Saito and Matsushita, 1967). Then the fact that several spectral peaks stand out, although the source wave apparently has a broad spectrum, leads us to conclude that the magnetosphere has an internal process to select certain spectral peaks and produce pc3, 4, and 5 waves with well-defined periodicities.

Another disturbance that is associated with the activation of pulsations is the expansion phase of the substorm. Pulsations thus activated include pi2 waves. The close relationship between the development of the expansion phase and of the pi2 activity is illustrated in Figure 108. Figure 108a displays the H- component records from the Canadian network of observatories (upper panel) and their high frequency version, which is produced by a filter with a corner frequency of 5×10^{-3} Hz (lower panel).

A sharp negative bay that signifies the onset of the expansion phase was recorded first at FTCH (66°) near 0645 and rapidly proceeded northward. The pi2 activity started at the same time, and as shown in Figure 108b, the occurrence of the pi2 onset in high latitudes is nearly concurrent with the arrival of the poleward border of the electrojet at a given station (Olson and Rostoker, 1975). The pi2 waves spread also to low latitudes in the night sector, and as they stand out in the relatively quiet background of the low-latitude field they provide very useful means for monitoring the onset of the expansion phase (Sakurai and Saito, 1976). It has been confirmed that the onset of the pi2 waves in low latitudes is closely associated with such basic expansion-phase signatures as the auroral break-up, the decrease in the magnetic field energy in the tail lobe, and the southward turning of the low-latitude magnetic field in the distant tail. A multiple onset of the expansion phase gives rise to a multiple onset of the low-latitude pi2 waves (Saito et al., 1976).

Note in the example of Figure 108 that the H component at most stations started to decrease gradually around ~ 0600. This decrease probably signified the growth phase of the substorm, and characteristically it was not associated with the pi2 activation.

The energy source of pi2 waves apparently lies in the tail field energy liberated by the reconnection process. It has been suggested that resonant oscillations are excited on night-side field lines when the tail field lines collapse and send the compressive wave toward the earth. One of the suggested sites of the resonant oscillation is the inner edge of the plasma sheet (Saito et al., 1976). Since the spectrum of the pi waves in high latitudes extends to the short period range and changes with the progress of the expansion phase, terms like pi1, piB, piC, and piP have been introduced to describe the detailed development of the pi phenomenon. Some of these waves may be produced at low altitudes in association with fluctuations in intensity and position of the precipitating auroral particles. In low latitudes pi2 waves have the form of a damped wave train with a fundamental period of $100 \sim 20$ s. The period is shorter when the Kp index is higher, and a suggestion has been made that the surface wave excited on the plasmapause is the nature of the low-latitude pi2 (Fukunishi and Hirasawa, 1970). The polarization of the pi2 waves has indeed been observed to reverse around the latitude of the plasmapause (Fukunishi, 1975). On this fundamental oscillation shorter period waves with periods less than 10 s are frequently superposed; the nature of these waves is not yet understood.

A close similarity has been noted between pi2 and pc4: not only are these two types of pulsations in the similar period range but they also show strong resemblance in the diurnal variation of the period, the

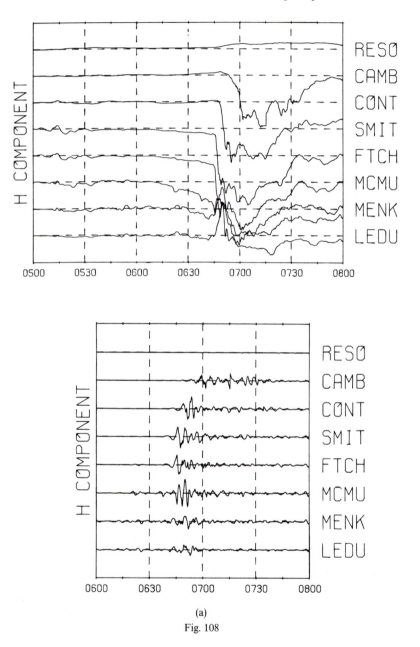

(a)

Fig. 108

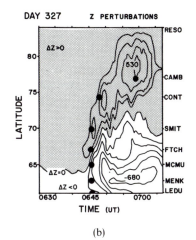

(b)

Fig. 108a,b. (a, facing page) Broad-band (top) and filtered (bottom) records of the magnetic field during a substorm event on 23 November 1970 recorded at the Canadian network of observatories. Interval between the baselines is 378γ (top) or 316γ (bottom). (b, above) Onset time of the pi2 wave (dot) and development of the electrojet shown by contours of the vertically downward perturbation ΔZ; $\Delta Z = 0$ delineates approximate center of the electrojet while the extrema identify the electrojet boundaries (Olson and Rostoker, 1975)

latitudinal variation of the amplitude, and the Kp dependence of the period. The difference, however, is that pc4 waves last longer while pi2 waves are damped, and this would be the reflection of the difference in the duration of the exciting agent (Fukunishi and Hirasawa, 1970).

V.4 Wave Generation by Wave–Particle Interactions

The solar-wind energy, which is transferred to the magnetosphere by the day-side reconnection process, emerges eventually in the inner magnetosphere in the form of the kinetic energy of the energetic particles. The radiation from these particles is an important source of the wave phenomena in the magnetosphere. Waves in the magnetic pulsation range are emitted from energetic protons by the cyclotron resonance process described by Eq. (111).

Cyclotron Instability

In order to assess the rate of the wave growth, let us utilize once again a one-dimensional model in which the uniform magnetic field $\boldsymbol{B}_0$ is directed in the positive z direction. In addition to the cold plasma of density N_1, hot protons of density N_3 are present, where $N_3 \ll N_1$ is assumed. Since the thermal velocity is no longer considered negligible as compared to V_A, it is necessary to introduce the distribution function. The distribution functions for the perturbed plasma are written as $F_j + f_j$ ($j = 1, 2, 3$ for cold protons, cold electrons, and hot protons, respectively), where F_j represents the distribution in the unperturbed state. The first order approximation to the Vlasov equation is

$$\frac{\partial f_j}{\partial t} + \boldsymbol{v} \cdot \frac{\partial f_j}{\partial \boldsymbol{r}} + \frac{e_j}{m_j}(\boldsymbol{v} \times \boldsymbol{B}_0) \cdot \frac{\partial f_j}{\partial \boldsymbol{v}} + \frac{e_j}{m_j}(\boldsymbol{E} + \boldsymbol{v} \times \boldsymbol{B}) \cdot \frac{\partial F_j}{\partial \boldsymbol{v}} = 0. \quad (143)$$

We assume that the perturbations f_j, $\boldsymbol{E}$, and $\boldsymbol{B}$ behave as $\exp\left(-i(\omega t - kz)\right)$ and set E_z to zero. Using the notations

$$(v_x, v_y, v_z) = (v_\perp \cos\theta, v_\perp \sin\theta, v_\parallel)$$

$$E_x = \mp iE_y \equiv E^\pm$$

$$f_j = f_j^\pm(\cos\theta \pm i \sin\theta)$$

$$\Omega_j = e_j B_0 / m_j$$

and writing $\boldsymbol{B}$ in terms of $\boldsymbol{E}$ (by using Faraday's law), we obtain

$$f_j^\pm = \frac{-ie_j E^\pm}{m_j(\omega - kv_\parallel \pm \Omega_j)}\left[\frac{\partial F_j}{\partial v_\perp}\left(1 - \frac{kv_\parallel}{\omega}\right) + \frac{kv_\perp}{\omega}\frac{\partial F_j}{\partial v_\parallel}\right]. \quad (144)$$

In writing E_x and E_y by $E^\pm$ we have assumed that the wave is circularly polarized; the upper sign corresponds to a wave with right-hand polarization with respect to $\boldsymbol{B}_0$ (namely, the fast-mode wave) and the lower sign to a left-hand (Alfven) wave. The current is related to f_j by

$$\boldsymbol{j} = \sum_j e_j \int \boldsymbol{v} f_j v_\perp \, dv_\perp \, d\theta \, dv_\parallel$$

and from Maxwell equations we have

$$i\omega\varepsilon_0\left(-\frac{c^2 k^2}{\omega^2} + 1\right)\boldsymbol{E} = \boldsymbol{j} \quad (145)$$

Hence the dispersion equation is

$$\frac{c^2 k^2}{\omega^2} = 1 + \frac{\pi}{\omega \varepsilon_0} \sum_j \frac{e_j^2}{m_j} \int \frac{dv_\perp \, dv_\parallel v_\perp^2}{\omega - kv_\parallel \pm \Omega_j} \cdot \left[\frac{\partial F_j}{\partial v_\perp} \left(1 - \frac{kv_\parallel}{\omega} \right) + \frac{kv_\perp}{\omega} \frac{\partial F_j}{\partial v_\parallel} \right] \quad (146)$$

Since the thermal spread of the velocity can be neglected for the cold plasma, Eq. (146) is simplified to

$$\frac{c^2 k^2}{\omega^2} = 1 - \frac{\omega_1^2}{\omega(\omega \pm \Omega_1)} - \frac{\omega_2^2}{\omega(\omega \pm \Omega_2)} + \pi \frac{\omega_3^2}{\omega} \int \frac{dv_\perp \, dv_\parallel v_\perp^2}{\omega - kv_\parallel \pm \Omega_3}$$

$$\cdot \left[\frac{\partial F_3}{\partial v_\perp} \left(1 - \frac{kv_\parallel}{\omega} \right) + \frac{kv_\perp}{\omega} \frac{\partial F_3}{\partial v_\parallel} \right] \quad (147)$$

where ω_j is defined by

$$\omega_j^2 = \frac{N_j e_j^2}{m_j \varepsilon_0}$$

and F_3 has been normalized (i.e., $\int F_3 \, d^3 v = 1$). The real part ω_R of ω can be obtained approximately from the first three terms. For integration of the fourth term we use the following formula

$$\int \frac{W(v_\parallel) \, dv_\parallel}{\omega - kv_\parallel \pm \Omega} = P \int \frac{W(v_\parallel) \, dv_\parallel}{\omega - kv_\parallel \pm \Omega} - \frac{i\pi}{|k|} W\left(\frac{\omega \pm \Omega}{k} \right)$$

where P indicates a principal value integral. We assume that the imaginary part ω_I of ω is small (i.e., $|\omega_I| \ll |\omega_R|$) and that k is real. When the distribution function of hot protons is a bi-Maxwellian distribution:

$$F_3 = \left(\frac{m}{2\pi\kappa T_\perp} \right) \left(\frac{m}{2\pi\kappa T_\parallel} \right)^{1/2} \exp \left[-\frac{mv_\perp^2}{2\kappa T_\perp} - \frac{mv_\parallel^2}{2\kappa T_\parallel} \right] \quad (148)$$

the growth rate ω_I is

$$\omega_I = \sqrt{\pi} \frac{N_3}{N_1} \frac{(\pm\Omega_3)(\omega_R \pm \Omega_3)^2}{k \left(\frac{2\kappa T_\parallel}{m} \right)^{1/2} (\omega_R \pm 2\Omega_3)\omega_R} [\pm\Omega_3 - (A + 1)(\omega_R \pm \Omega_3)]$$

$$\cdot \exp \left(-\frac{m}{2\kappa T_\parallel} \frac{(\omega_R \pm \Omega_3)^2}{k^2} \right) \quad (149)$$

where A represents the temperature anisotropy:

$$A = \frac{T_\perp}{T_\parallel} - 1.$$

Thus in the present system ω has an imaginary part due to the cyclotron resonance of hot protons whose parallel velocity $v_\parallel$ is high enough to satisfy $\omega - kv_\parallel = \mp \Omega_3$. Without loss of generality we can assume that $\omega_R > 0$. Then for the fast-mode wave (that corresponds to the upper sign) the preceding condition means $v_\parallel > \left(\dfrac{\omega}{k}\right)$, namely, that the wave resonates with protons that move faster than the wave. For the Alfven-mode wave (corresponding to the lower sign) the resonance occurs with protons that move opposite to the wave (i.e., $kv_\parallel < 0$) since the existence of the Alfven-mode is limited to frequencies below the proton gyrofrequency (i.e., $\omega < \Omega_3$). In the frame of reference moving with resonant protons, therefore, the wave field of both modes is seen to rotate with the proton gyrofrequency in the sense of the proton gyration (i.e., with the left-hand sense with respect to B). The resonance causes the wave to grow when ω_I is positive, namely, when

$$-A > \frac{1}{\dfrac{\Omega_3}{\omega} + 1} \qquad \text{for fast mode} \tag{150a}$$

$$A > \frac{1}{\dfrac{\Omega_3}{\omega} - 1} \qquad \text{for Alfven mode} \tag{150b}$$

where ω is used instead of ω_R. Hence the fast-mode wave is excited if $T_\parallel > T_\perp$ and the Alfven-mode if $T_\perp > T_\parallel$. Hereafter we shall concentrate on the latter case because in the magnetosphere $T_\perp > T_\parallel$ is the more prevalent condition. Inequality (150b) is equivalent to

$$\omega < \frac{A}{A + 1} \Omega_3 \tag{151}$$

and hence there is an upper limit to the frequency that can be excited. The ratio of ω_I to Ω_3 is plotted in Figure 109 for different values of the anisotropy ratio A (Fig. 109a) and of the mean thermal speed ($U_\parallel \equiv (2\kappa T_\parallel/m)^{1/2}$) parallel to B_0 (Fig. 109b) against the normalized frequency ω/Ω_3. It is evident that the wave grows faster when the degree of the anisotropy is greater and when the parallel temperature of hot protons is higher.

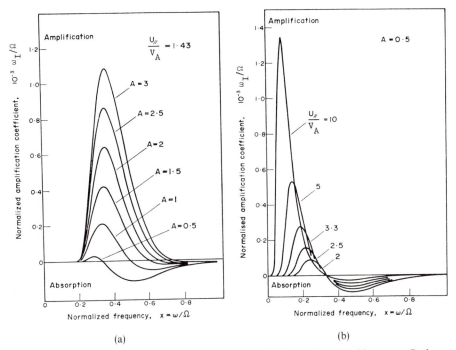

Fig. 109a,b. Growth rate of the proton cyclotron instability as a function of frequency. Both quantities are shown normalized by the proton gyrofrequency $\Omega(\equiv\Omega_3)$. (a) Dependence on the anisotropy ratio A, and (b) dependence on the ratio between the mean thermal speed and the Alfven speed (Gendrin et al., 1971)

Short-Period Magnetic Pulsations

The proton cyclotron resonance theory of wave excitation as summarized above has been developed to explain the origin of waves of pc1 and pi1 types (Kennel and Petschek, 1966; Cornwall, 1965). If we take the equatorial region as the likely place of the wave excitation because the energetic particle density maximizes there, the excitation of the waves in these period ranges can be expected in the outer magnetosphere since the proton gyrofrequency falls in the $1 \sim 20$ Hz range beyond $L \sim 5$. The short-period magnetic pulsations have been classified into several subsets, because there are differences in the manner with which the wave frequency varies with time. Two distinct examples are displayed in Figure 110; in the first example the frequency rises sharply with time and the structure is repeated periodically, while in the second example the spectrum is broad and the frequency stays nearly at the same level (before ~ 0230) or rises gradually with time (between ~ 0230 and ~ 0300). Waves of the first type

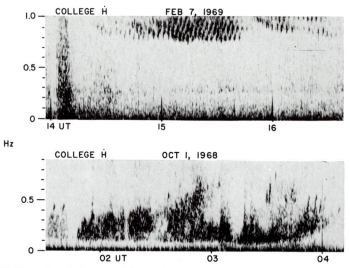

Fig. 110. Examples of pcl waves of the pearl-type (upper panel) and of the IPDP-type (lower pannel, middle part) (Heacock and Akasofu, 1973)

are frequently called 'pearls' because the periodic repetition of the wave packet gives a pearl necklace-like appearance to the amplitude-time display of the wave. The waves with the gradual rise in frequency are often referred to as IPDP (interval of pulsations of diminishing periods) events. The local time dependence of the occurrences of various types of short-period pulsations at the auroral-zone station is summarized schematically in Figure 111 (Kokubun, 1970). (In this Figure the spectral structure is shown on an expanded time scale and hence should not be measured by the scaling of the abscissa.) As the Figure shows, there are several more

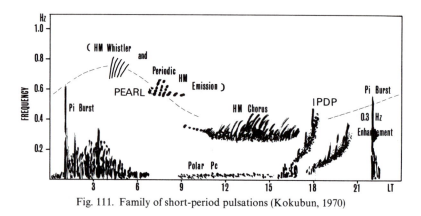

Fig. 111. Family of short-period pulsations (Kokubun, 1970)

subsets of short-period pulsations in addition to pearl and IPDP, but it is these two types of waves with which the proton cyclotron resonance theory has been extensively compared.

Periodic rising-frequency structures that characterize the spectrum of pearls have been found to appear alternately in opposite hemispheres; the signal appears in the northern (southern) hemisphere during the interval between the appearances of the signals in the southern (northern) hemisphere. This is illustrated in Figure 112 where observations at Great Whale River ($67°$) and at Byrd Station ($-71°$) are compared. In the top panel the sonagrams (i.e., frequency-time displays) from these two stations are superposed, and the alternate appearances of the rising-frequency structures in opposite hemispheres are demonstrated. In the lower panel the same feature is shown by a comparison of the amplitude records. This observation suggests that the periodic structure of the pearl waves reflects the bouncing of the wave packet along the field line between opposite hemispheres (Tepley, 1964). Since the rising-frequency structure tends to be repeated many times, although a substantial fraction of its energy should be lost from the wave packet at each reflection at the ionosphere, the energy of the packet should be reinforced each time when the packet passes through the amplifying region around the equator.

The rising tone of the wave packet of the pearl wave has been attributed to the effect of the dispersion. The idea was stimulated by the observation that the rate of the frequency-rise often shows a systematic decrease with the repetition of the signal (Obayashi, 1965; Jacobs and Watanabe, 1964); the sonagrams of these pearls are said to have a fanning structure. Since we are interested in frequencies much less than the electron gyrofrequency ($\omega \ll |\Omega_2|$), the real part of Eq. (147) simplifies to

$$\frac{c^2 k^2}{\omega^2} \doteq -\frac{\omega_1^2}{\omega(\omega - \Omega_1)} + \frac{\omega_2^2}{\omega\Omega_2} = \frac{c^2/V_A^2}{1 - \dfrac{\omega}{\Omega_1}}. \tag{152}$$

Hence the group velocity of the wave packet is

$$V_g = V_A(1 - x)^{3/2}(1 - x/2)^{-1} \tag{153}$$

where $x = \omega/\Omega_1$. Let us consider a case in which the wave was generated at $t = 0$ simultaneously over a certain frequency range around ω_0 and took time $t(\omega)$ to travel the distance z. Since $t(\omega) = z/V_g(\omega)$, we have

$$\frac{dt}{d\omega} = -\frac{z}{V_g^2}\frac{dV_g}{d\omega} = -\frac{t}{V_g}\frac{dV_g}{d\omega}. \tag{154}$$

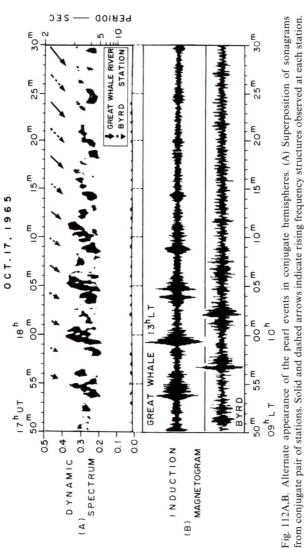

Fig. 112A,B. Alternate appearance of the pearl events in conjugate hemispheres. (A) Superposition of sonagrams from conjugate pair of stations. Solid and dashed arrows indicate rising frequency structures observed at each station. (B) Amplitude vs. time records of the induction magnetogram corresponding to (A). (Saito, 1969)

It is assumed in the foregoing that the medium is uniform, and V_{g} and its derivative will be represented by their values at the equator. This assumption has been found to be reasonable since it is at the equatorial region where x is not negligibly small as compared to 1 and the wave packet spends most of its travel time. Combining Eqs. (153) and (154) one obtains

$$t \frac{d\omega}{dt} = \Omega_1 (1 - x) \left(1 - \frac{x}{2}\right) \left(1 - \frac{x}{4}\right)^{-1}. \tag{155}$$

Thus the dispersion causes the rate $d\omega/dt$ of the frequency rise to decrease as t^{-1}. Conversely, when the observed $dt/d\omega$ is a linear function of t, we can estimate the proton gyrofrequency Ω_1 at the equator by substituting the observed values of ω_0 and $t \dfrac{d\omega}{dt}$ (at $\omega = \omega_0$) in Eq. (155), and hence identify the L value of the field line on which the wave packet has propagated (Gendrin et al., 1971).

As a matter of fact, not all the pearl events have the fan structure when seen in sonagrams; there are cases where the rising-frequency structures are approximately parallel. To distinguish between the two types of pearls, those with the fan structure are often referred to as HM whistlers while those with the parallel structure are called HM emissions. As a possible cause of this difference Gendrin et al. (1971) noted that Eq. (154) should in fact be written as

$$\frac{dt}{d\omega} \doteq - \frac{\left\{\left(\dfrac{d^2\omega_1}{dk^2}\right)^2 + \left(\dfrac{dV_g}{dk}\right)^2\right\} t}{V_g^2 \dfrac{dV_g}{dk}}$$

when $\left|\dfrac{d^2\omega_1}{dk^2}\right|$ is large. Hence if ω_1 is large enough to make $(d^2\omega_1/dk^2)^2 \gg (dV_g/dk)^2$ and varies with time as t^{-1}, due, for example, to the relaxation of the distribution function associated with the wave excitation, $dt/d\omega$ can stay approximately constant and the parallel structure may be produced.

Despite the foregoing, past studies have been concerned mostly with the pearl events with fan structure. Once Ω_1 and hence L-value of the path is derived by using the dispersion model, the period τ of the repetition of the wave packet can be used to yield information on the equatorial plasma density. τ is given by

$$\tau = 2 \int \frac{dz}{V_g} \tag{156}$$

where the path of the integration is the field line between the two conjugate ionospheres. The integration can be performed if some assumption is made about the variation of the ambient plasma density N_1 along the field line, and then the density at the equator can be derived from the observed value of τ. Information on the temperature of the hot protons, on the other hand, can be obtained from the frequency ω_0 where the wave is strongest, since ω_0 can be compared with the frequency at which ω_1 maximizes. This frequency depends both on A and $U_\parallel/V_A$, but Figure 109 shows that the dependence on the latter parameter is more pronounced. Gendrin et al. (1971) applied the preceding method to about 40 cases of the pearl events and found that the anisotropy factor A is of the order of 1 to 2 and that the mean thermal velocity $U_\parallel$ parallel to $\boldsymbol{B}_0$ lies in the following range.

$$2 < U_\parallel/V_A < 10 \text{ for } L \sim 4$$

$$0.3 < U_\parallel/V_A < 2 \text{ for } L \sim 5.$$

Assuming that N_1 is proportional to B_0 and setting A equal to 2, they found that the energy $E_3 = \frac{m_3}{2}(A + 2)U_\parallel^2$ of the hot protons varies with L as $E_3 = 11L^{-3}$ MeV.

The process of the cyclotron resonance modifies the distribution function of hot protons. This in turn gives rise to systematic changes in the spectral shape of the waves produced by the resonance (Cocke and Cornwall, 1967; and others). The consideration of this nonlinear effect is apparently needed to understand the variety of spectral structures of short-period pulsations, but we shall not pursue the subject due to the lack of space.

The dispersion analysis places the wave packet of the pearl events mostly on field lines at L of $4 \sim 8$ (Gendrin et al., 1971; also Kenney et al., 1968 and others based on slightly different techniques). This is the region that encompasses the average plasmapause position. By sorting out the occurrence probability of the pc1 waves according to local time and Kp, Roth and Orr (1975) have noted that the probability is high when the observing site is located inside the statistically determined position of the plasmapause for the given LT and Kp. As for the relation to the large-scale disturbance phenomena, Heacock and Kivinen (1972) have reported that the pearl type pc1 waves have a tendency to occur during the late epoch of ring current recoveries. Since the growth rate of the ion cyclotron instability increases when V_A is reduced (cf. Fig. 109b), the latter observation has been interpreted to suggest the following: pearls are produced when the ionospheric plasma is replenished to the expanded outer plasmasphere and reduces V_A in the region where the fresh population of energetic protons has been introduced from the magnetotail during the storm's main phase.

In any event, the present model of the pearl-type pc1 waves implies that the wave packet is produced only within a limited range of L in the form of a packet whose duration, as seen at any given frequency, is much less than the bounce period; otherwise the distinctly separated periodic structures would not be produced in the sonagram. This characteristic does not seem to have been explained yet.

Short period pulsations of the IPDP type, on the other hand, are the evening-side phenomena that follow the onset of the substorm expansion phase with a short (<1 h) delay (Fukunishi, 1969; Heacock, 1971). These pulsations are therefore considered to be the more immediate product of the hot particles injected from the magnetotail. Reference to Figure 109 suggests that there are at least two possible ways to explain the diminishing period by the cyclotron resonance theory. The first is the deeper penetration of the protons, since ω of the most unstable wave is nearly proportional to the local proton gyrofrequency. The dawn-to-dusk electric field that is required to explain the frequency shift by this idea is about 1 mV/m, which is in reasonable agreement with the values estimated by other means (cf. Chapter IV) (Troitskaya et al., 1968). The second way is the westward drift of energetic protons from the near-midnight meridian; protons with the lower energy (above several keV, though) reach the evening sector at the later epoch since their azimuthal drift speed is slower and produces the higher-frequency wave since ω_I maximizes at higher ω when $U_\parallel/V_A$ is lower (Fukunishi, 1969). ($U_\parallel \propto U_\perp$ is implied in this argument.) Heacock et al. (1976) have found that the rate of the frequency shift of IPDP is positively correlated with the AE index, and suggested that this observation is consistent with the first hypothesis since AE can be regarded as a rough measure of the dawn-to-dusk electric field on the night-side. Fukunishi (1969), on the other hand, showed that the rate df/dt of the frequency-rise decreases with the increase in the delay of the local IPDP onset from the expansion-phase onset time, and he presented this as evidence in favor of the second hypothesis. Unlike the pearls, fine structures within the IPDP events do not show any correlation at conjugate points, indicating that the wave amplification process is chaotic and occurs at many points (Heacock et al., 1976).

Sudden compressions of the magnetosphere (as evidenced by occurrences of positive SIs) can exert influence on the activity of short-period pulsations. At the auroral zone the influence is observed most frequently in the noon–evening sector. If the wave activity is already in progress when the compression occurs, the compression tends to increase the average frequency and bandwidth of the wave and produce an irregular series of rising tones that lasts for 20 min to 2 h. If short-period pulsations are absent in the preceding interval a burst-like enhancement of the activity is observed for several minutes. Time delays of $0.5 \sim 3$ min are noted between SSC (positive SI) and the activation of the wave (Kokubun and Oguti, 1968). The enhancement of the wave activity is likely to be due to the enhancement in the anisotropy ratio caused by the adiabatic compression (cf. Fig. 109a). A formula has been derived to estimate the L coordinate of the wave packet from $\Delta\omega$ and the magnitude ΔH of the SI field on the ground (Troitskaya and Gulelmi, 1967). The observed time delay probably reflects the rise time of SI.

Interaction of Particles with Long-Period Waves

The resonant interaction between waves and particles can act also as a source of long-period waves. One of the possible mechanisms is the drift instability. When the medium is nonuniform in the x direction, the unperturbed distribution function becomes a function of $x + \frac{1}{2}\varepsilon x^2 + v_y/\Omega$ if the unperturbed magnetic field is directed in the z-direction and its x-dependence is expressed as $B_0(1 + \varepsilon x)$. Hasegawa (1969) derived the dispersion equation by using the following as the unperturbed distribution function of the hot component

$$
F = N_0 \left(\frac{1}{\pi U_\perp}\right) \left(\frac{1}{\pi U_\parallel}\right)^{1/2} \left\{ 1 - \kappa \left(x + \frac{1}{2}\varepsilon x^2 + \frac{v_y}{\Omega}\right)\right\}
$$

$$
\cdot \exp\left\{ -\frac{v_\parallel^2}{U_\parallel^2} - \frac{v_\perp^2}{U_\perp^2}\right\}. \tag{157}
$$

The presence of the cold, but dense, component is also assumed, and it was found that the real part of ω is given by the ion drift frequency ω_i^*:

$$
\omega_i^* = k_y v_{d,i} = \frac{k_y \kappa U_{\parallel,i}^2}{\Omega_i}
$$

where $v_{d,i}$ is the drift velocity of ions in the presence of the density gradient. The perpendicular wave number k_y that corresponds to the maximum growth rate is

$$k_y = \frac{(2\Delta)^{1/2}}{3} \frac{\Omega_i}{U_{\perp,i}} \tag{158}$$

where

$$\Delta = \frac{3}{4} \left\{ \frac{N_i \kappa T_{\perp,i}}{\left(\frac{B_0^2}{2\mu_0}\right)} \left(\frac{T_{\perp,i}}{T_{\|,i}} - 1\right) - 1 \right\}.$$

The instability, called drift mirror instability, takes place when the ion temperature anisotropy is large enough to satisfy $\Delta > 0$. Re (ω) that corresponds to the maximum growth rate is

$$\mathrm{Re}\,(\omega) = \frac{\kappa U_{\|,i}}{3} \left(\frac{T_{\|,i}}{T_{\perp,i}}\right)^{1/2} (2\Delta)^{1/2}. \tag{159}$$

(The electron drift wave frequency was ignored in the foregoing derivation.) The perturbations in pressure and magnetic field are related by

$$p_{\perp,i} \sim 2P_{\perp,i} \left(1 - \frac{T_{\perp,i}}{T_{\|,i}}\right) \frac{B_z}{B_0}$$

$$p_{\|,i} = P_{\|,i} \left(1 - \frac{T_{\perp,i}}{T_{\|,i}}\right) \frac{B_z}{B_0} \tag{160}$$

where P represents the pressure in the unperturbed state.

This theory was compared with a distinct case of quasi-periodic oscillations in field and particle flux observed by Explorer 26 during the main phase of a magnetic storm. As shown in Figure 113, the oscillations started soon after the satellite entered the region of energetic protons in which the ring current was flowing. The local time of the satellite was about 1300. Characteristically, the flux of protons having a nearly perpendicular pitch angle fluctuated out of phase with magnetic-field variations, while the protons having a much smaller pitch angle showed little sign of a coherent fluctuation (Brown et al., 1968). These relationships between field and particle fluctuations are consistent with Eq. (160), but it could not be confirmed that the temperature anisotropy was strong enough to satisfy the instability condition $\Delta > 0$ (Lanzerotti et al., 1969).

The excitation of long-period pulsations by wave–particle interactions that take place in the ring-current protons has been inferred also from

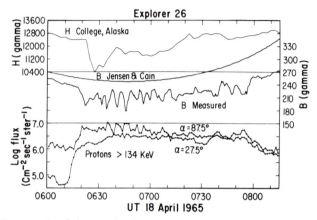

Fig. 113. An example of the correlated, quasi-periodic fluctuations in the magnetic field and the energetic proton flux observed in the equatorial region at L ∼ 5. α means the particle pitch angle with respect to the magnetic field. H-component magnetogram from an auroral-zone station is shown for reference in the top panel. (Brown et al., 1968)

satellite observations of pc5 waves during the main phase of a storm. These observations were made by the synchronous equatorial satellite ATS-1, and storm-time pc5 waves were detected only when the satellite was orbiting through the afternoon sector. Although simultaneous plasma observations were not performed, the occurrence of waves during the main phase and their confinement to the afternoon sector make it likely that these waves were generated in the high-β asymmetric ring-current protons injected from the magnetotail. The wave events lasted usually for about one hour, and the spectral characteristics such as the auto spectra, ellipticity, orientation, etc., were kept nearly the same throughout each individual event. The ellipticity was pronounced (with a mean value of less than 0.1) and the major axis of the polarization ellipse was very close to the meridian plane (within 19° from that plane) (Barfield et al., 1972; Barfield and McPherron, 1972). According to these authors the near alignment of the major axis to the meridian plane favors the interpretation that the observed waves are Alfven waves, rather than drift waves. The bounce–resonant interaction, which incorporates the resonant coupling between particles bouncing between mirror points and Alfven waves bouncing between ionospheric reflection heights, has been suggested as a possible means for exciting these long-period waves in the ring-current protons.

The wave that grows at lower altitudes from the beam of precipitating particles is usually considered to be in the VLF range, since at low altitudes Ω is high. If it happens, however, that the charge carried by the beam is not completely neutralized, instability can occur in the hydro-magnetic range. The frequency of the excited wave, which is proportional to the density of the nonneutralized beam and the gyrofrequency of the ions, was found to fall in $10^{-2} \sim 10^{-3}\,\mathrm{s}^{-1}$ range for parameters of the auroral-zone precipitation during substorms. Since there are indications that the auroral precipitations are not entirely devoid of the space charge, this process may be responsible for some of the pulsations in the pi1 and pi2 ranges (Nishida, 1964a; Kimura and Matsumoto, 1968).

More often, however, oscillations observed in the flux of precipitating electrons (as oscillations in auroral luminosity, cosmic ray absorption or bremsstrahlung X-ray intensity) in association with magnetic pulsations are likely to be due to modulation of the particle pitch angle by hydro-magnetic waves. (Precipitating electrons may influence the amplitude and polarization of pulsations observed on the ground by modulating the ionospheric conductivity, but the periodic oscillation of the precipitating flux should have its origin in the magnetospheric process and hence in hydromagnetic waves.) On these occasions the oscillations observed in the precipitating electron flux may be regarded as secondary phenomena, but they nevertheless represent an important feature as they give vital information on the spatial structure of the magnetospheric oscillations when data are obtained on the spatial pattern of the auroral luminosity fluctuations (so-called auroral coruscations) that are correlated with magnetic pulsations. According to Oguti and Watanabe (1976) some of the auroral-zone pulsations with periods of a few seconds to several tens of seconds that often occur in the dawn sector during the substorm recovery phase are related to a quasi-periodic poleward propagation of the aurora. They have interpreted that these waves are fast-mode waves that are produced quasi-periodically around the inner edge of the injected ring-current protons and emitted outward.

Thus, correlated studies of magnetospheric particles and magnetic pulsations represent a very important and promising field of research that will attract much attention in coming years.

V.5 Ionospheric and Atmospheric Modification of Hydromagnetic Waves

Before reaching the ground to be detected as magnetic pulsations, hydro-magnetic waves generated in the magnetosphere have to pass through the ionosphere and the atmosphere. In the ionosphere, part of the wave energy is dissipated and the axis of the polarization ellipse is rotated due to the effect of collisions between ions and neutral particles. In the neutral atmosphere that occupies the space between the ionosphere and the ground, the wave field is endowed with some specific characteristics due to the absence of current and charge. In boundary regions where the nature of the medium changes sharply, both reflection and mode coupling take place. All these factors have to be taken into consideration in order to relate magnetic pulsations with the wave phenomena in the magnetosphere.

Ionospheric Screening Effect

We shall derive the basic features of the ionospheric–atmospheric modification by using a very simplified model. The relevance of the result will be confirmed later by comparison with exact numerical calculations. For simplicity the space is considered to be one-dimensional and is divided into four uniform layers: magnetosphere, ionosphere, atmosphere, and the earth. The ambient magnetic field is assumed to be directed vertically downward, namely, opposite to the z axis. The perturbation field is assumed to depend on time and on a horizontal coordinate as $\exp(i(mx - \omega t))$. In the magnetosphere where the ion gyrofrequency exceeds the ion-neutral collision frequency, the wave field satisfies Eqs. (114) and (115), which can be written in the present geometry as

$$
\left(\frac{\omega^2}{V_A^2} + \frac{\partial^2}{\partial z^2}\right) E_x = 0
$$
$$
\left(\frac{\omega^2}{V_A^2} + \frac{\partial^2}{\partial z^2} - m^2\right) E_y = 0. \tag{161}
$$

The two equations give dispersion relations for the Alfven mode and the fast mode, respectively. The wave field in the magnetosphere can hence be written as

$$
B_x = (\alpha_{1I} e^{-i\mu_1(z-d)} + \alpha_{1R} e^{i\mu_1(z-d)}) e^{i(mx-\omega t)}
$$
$$
E_y = \frac{\omega}{\mu_1} (\alpha_{1I} e^{-i\mu_1(z-d)} - \alpha_{1R} e^{i\mu_1(z-d)}) e^{i(mx-\omega t)} \tag{162}
$$

for the fast mode designated by suffix 1. Here $\mu_1^2 = \omega^2/V_A^2 - m^2$. The suffices I and R designate waves that propagate downward and upward, respectively, and $z = d$ is the altitude of the ionosphere which is represented by an infinitely thin layer. For the Alfven mode designated by suffix 2, we have

$$B_y = (\alpha_{21}e^{-i\mu_2(z-d)} + \alpha_{2R}e^{i\mu_2(z-d)})e^{i(mx-\omega t)}$$

$$E_x = -\frac{\omega}{\mu_2}(\alpha_{21}e^{-i\mu_2(z-d)} - \alpha_{2R}e^{i\mu_2(z-d)})e^{i(mx-\omega t)} \tag{163}$$

where $\mu_2 = |\omega/V_A|$.

In the neutral atmosphere where no current flows, the perturbation field should satisfy

$$\left(\frac{\omega^2}{c^2} + \frac{\partial^2}{\partial z^2} - m^2\right)E = 0 \tag{164}$$

Now while m should be greater than 10^{-3} km^{-1} as the perturbation field is seen to vary in distances shorter than the earth's radius, (ω/c) is less than 10^{-5} km^{-1} for low-frequency waves with $\omega < 3$ s^{-1}. Hence the wave should be evanescent in the vertical direction if it propagates horizontally with a real m, and hence it can be expressed as

$$B_x = \beta_1(e^{|m|z} + e^{-|m|z})e^{i(mx-\omega t)}$$

$$E_y = -\frac{i\omega}{|m|}\beta_1(e^{|m|z} - e^{-|m|z})e^{i(mx-\omega t)} \tag{165}$$

$$B_y = 0$$

$$E_x = \frac{i\omega}{|m|}\beta_2(e^{|m|z} - e^{-|m|z})e^{i(mx-\omega t)} \tag{166}$$

where E_x and E_y are assumed to disappear on the ground ($z = 0$) due to the very high conductivity of the earth. The important property $B_y = 0$ is derived from

$$\mu_0 j_z = \frac{\partial B_y}{\partial x} - \frac{\partial B_x}{\partial y} = \frac{\partial B_y}{\partial x} = 0.$$

At the ionosphere the horizontal electric field is continuous but the magnetic field is changed by the current flowing therein. These boundary conditions can be written as

$$-\frac{i}{|m|}\beta_1(e^{|m|d} - e^{-|m|d}) = \frac{1}{\mu_1}(\alpha_{1I} - \alpha_{1R})$$

$$\frac{i}{|m|}\beta_2(e^{|m|d} - e^{-|m|d}) = -\frac{1}{\mu_2}(\alpha_{2I} - \alpha_{2R})$$

$$\mu_0\left[-\Sigma_P\frac{\omega}{\mu_2}(\alpha_{2I} - \alpha_{2R}) + \Sigma_H\frac{\omega}{\mu_1}(\alpha_{1I} - \alpha_{1R})\right]$$
$$= -(\alpha_{2I} + \alpha_{2R})$$

$$\mu_0\left[\Sigma_P\frac{\omega}{\mu_1}(\alpha_{1I} - \alpha_{1R}) + \Sigma_H\frac{\omega}{\mu_2}(\alpha_{2I} - \alpha_{2R})\right]$$
$$= (\alpha_{1I} + \alpha_{1R}) - \beta_1(e^{|m|d} + e^{-|m|d}). \quad (167)$$

α_{1I} and α_{2I} represent amplitudes of fast-mode and Alfven-mode waves incident on the ionosphere and Eqs. (167) give amplitudes of reflected and transmitted waves in terms of these. Amplitude β_1 of the transmitted magnetic perturbation field is

$$\beta_1 = \frac{-\dfrac{2}{\mu_1}\left\{\left(1 + \dfrac{S_P\omega}{\mu_2}\right)\alpha_{1I} - \dfrac{S_H\omega}{\mu_2}\alpha_{2I}\right\}}{-\left(1 + \dfrac{S_P\omega}{\mu_2}\right)\varDelta_+ + \dfrac{S_H\omega}{\mu_1}\dfrac{S_H\omega}{\mu_2}\varDelta_- + \left(1 + \dfrac{S_P\omega}{\mu_1}\right)\left(1 + \dfrac{S_P\omega}{\mu_2}\right)\varDelta_-}$$

$$(168)$$

and the amplitude α_{2R} of the reflected Alfven-mode wave is

$$\alpha_{2R} = \frac{-2\dfrac{S_H\omega}{\mu_1}\varDelta_-\alpha_{1I} + \left\{\begin{array}{l}\left(1 - \dfrac{S_P\omega}{\mu_2}\right)\varDelta_+ + \dfrac{S_H\omega}{\mu_1}\dfrac{S_H\omega}{\mu_2}\varDelta_- \\[2mm] \qquad - \left(1 + \dfrac{S_P\omega}{\mu_1}\right)\left(1 - \dfrac{S_P\omega}{\mu_2}\right)\varDelta_-\end{array}\right\}\alpha_{2I}}{-\left(1 + \dfrac{S_P\omega}{\mu_2}\right)\varDelta_+ + \dfrac{S_H\omega}{\mu_1}\dfrac{S_H\omega}{\mu_2}\varDelta_- + \left(1 + \dfrac{S_P\omega}{\mu_1}\right)\left(1 + \dfrac{S_P\omega}{\mu_2}\right)\varDelta_-}$$

$$(169)$$

where the following notations have been used:

$$S_P = \mu_0 \Sigma_P, \qquad S_H = \mu_0 \Sigma_H.$$

$$\Delta_+ = \frac{1}{\mu_1}(e^{|m|d} + e^{-|m|d}), \qquad \Delta_- = \frac{i}{|m|}(e^{|m|d} - e^{-|m|d}).$$

As an illustration, let us consider a situation where $\omega = 0.1/s$ (namely, $T \sim 60$ s) and $|m| = (25)^{-1}/$km (namely, horizontal wavelength ~ 150 km). Since V_A is $10^{3\sim4}$ km/s in the lowermost magnetosphere, $\mu_1^2 \simeq -m^2$ in this case and the fast mode becomes an evanescent wave. Then the fast-mode wave with suffix 1I represents a wave whose amplitude increases upward if we set $\mu_1 = i|m|$. The electric conductivity in the middle latitude ionosphere is such that during the day $\Sigma_P \sim 10$ mho ($\sim 10^{13}$ esu) and hence $S_P\omega/\mu_2 \gg 1$ but $S_H\omega/\mu_2 \lesssim 1$. Also the inequality $S_P\omega/\mu_1 \ll 1$ holds as $|\mu_1| \simeq |m| \gg |\omega/V_A| = |\mu_2|$. Hence we can approximate Eqs. (168) and (169) as

$$\beta_1 \simeq \left(\alpha_{11} - \frac{S_H}{S_P}\alpha_{21}\right)e^{-|m|d} \tag{170a}$$

$$\alpha_{2R} \simeq \frac{i\mu_2 S_H}{|m|S_P}\alpha_{11} + \left(1 - \frac{\mu_2}{S_P\omega}\right)\alpha_{21}. \tag{170b}$$

The numerical solution of the same problem, in which the vertical structure of the ionosphere is fully taken into account, is shown in Figure 114, and we can see that the basic features of this solution are consistent with Eqs. (170). In this Figure the magnitudes of the magnetic and the electric perturbation fields are plotted on the left-hand side against altitude, and their relative phase angles are shown on the right-hand side (Hughes and Southwood, 1976a). Only the Alfven-mode wave is considered to be incident. The first point to note is that while the incident magnetic perturbation field is directed perpendicularly to the wave vector, the magnetic perturbation received on the ground is parallel (or, anti-parallel) to the wave vector in the horizontal plane. (Namely, B_x is produced on the ground from the incidence of B_y.) This is due to the $j_z = 0$ condition imposed in the neutral atmosphere, and the rotation of the B vector occurs in the ionosphere by virtue of the Hall current. B_x on the ground is related to B_y incident on the ionosphere by the first part of Eq. (170):

$$B_{xG} = -2\frac{S_H}{S_P}e^{-|m|d}B_{yM} \tag{171}$$

MAGNITUDE   PHASE

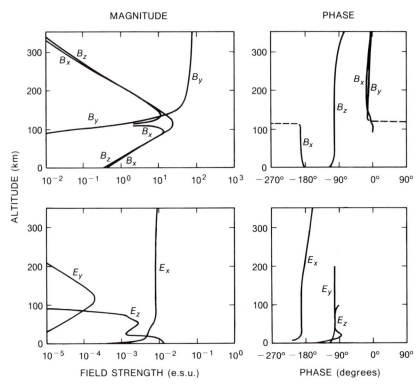

ALTITUDE (km)

FIELD STRENGTH (e.s.u.)   PHASE (degrees)

Fig. 114. Vertical profiles of the magnetic and the electric perturbation fields: a model calculation using the daytime, sunspot maximum ionosphere model (Hughes and Southwood, 1976a)

where factor 2 represents the effect of the ground reflection. Since the z-axis is taken vertically upward, this means that the magnetic perturbation vector is rotated through 90° in the counterclockwise direction when viewed toward the ground. In the southern hemisphere the minus sign in front of the right-hand side is replaced by a plus sign and the rotation of the magnetic perturbation occurs in the clockwise direction when viewed toward the ground. Due to the $\exp(-|m|d)$ factor, waves having a shorter wavelength suffer a greater attenuation before reaching the ground.

The height profiles of B_x and B_z in Figure 114 are sharply peaked around the height of the maximum ionospheric conductivity. This is because at lower altitudes they behave as $\exp(|m|z)$, while at higher altitudes the fast-mode decays as $\exp(-|m|z)$ in the present circumstance. The incident Alfvén wave is almost perfectly reflected when $S_P\omega/\mu_2 \gg 1$, because $\alpha_{2R} \simeq \alpha_{2I}$ then. This condition applies in the daytime, but during the night $S_P\omega/\mu_2 \sim 1$ and there is significant dissipation of the wave energy during the passage through the ionosphere. This has been suggested as a possible cause of the damping of the night-time pi pulsations (Hughes and Southwood, 1976a).

When the wave incident on the ionosphere consists only of the fast mode, Eq. (170a) becomes $B_{xG} = 2B_{xM}e^{-|m|d}$ and the magnetic perturbation vector is not rotated during its transmission through the ionosphere.

A preliminary attempt has been made to combine the preceding theory with the result of the hydromagnetic resonance theory described in Section V.2. Figure 115 illustrates the resulting relationship between the magnetospheric and the ground perturbation fields (Hughes and Southwood, 1976b). For this example the azimuthal wave number of $1/1000 \text{ km}^{-1}$, angular frequency of 0.1 s^{-1}, and the latitudinal scale length of the resonant region $\varepsilon = 10$ km are adopted, and the morning side is considered. The continuous lines in the Figure show magnetospheric parameters; the wave energy is transmitted toward the $-x$ direction and the resonant condition is met at $x = 0$. Ground signal parameters are shown by dashed lines. The top panel shows that the sharp resonant peak in the transverse oscillation B_y in the magnetosphere produces on the ground a corresponding peak in B_x. The amplitude of the peak, however, is much lower since the $\exp(-|m|d)$ factor in the β_1/α_{21} ratio acts to reduce the amplitude of the sharp structure. The azimuth (tilt) of the horizontal polarization ellipse, shown in the bottom right-hand panel, shifts through right angles from NE in the magnetosphere to NW on the ground due to the ionospheric rotation of the magnetic perturbation vector. The ellipticity of the horizontal polarization ellipse is shown in the bottom left-hand panel. As we have seen in Section V. 2, the ellipticity in the magnetosphere is dependent on the horizontal gradient of the wave amplitude and changes sign at the resonant peak, but on the ground the Figure shows that the ellipticity has the same sign throughout the resonant region. This result is annoying since observationally the reversal of the polarization is a well-established feature and has been employed to identify the resonance region.

There remains a possibility, however, that the absence of the polarization reversal in the preceding calculation is due to the oversimplicity of the model. In Hughes and Southwood (1976b) the horizontal distribution of the perturbation field is expanded into the Fourier series. This procedure would not be pertinent, because in the resonance model the field distribution in the direction of the nonuniformity of the medium is evanescent, rather than propagating. The resonance model supposes, instead, that the wave vector is real in the vertical direction and the incident wave involves the fast mode whose perturbation magnetic vector

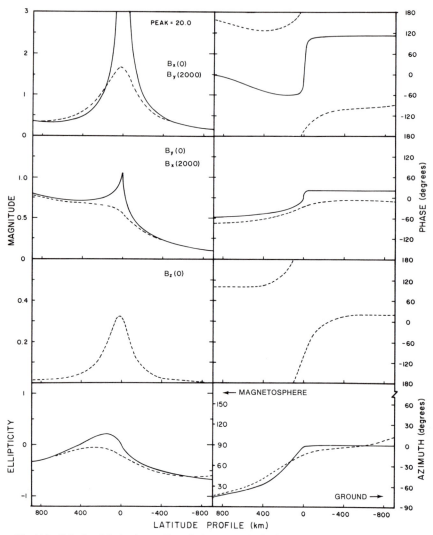

Fig. 115. Calculated latitude profiles of the magnetic field components and polarization characteristics at the ground level (–––––) and at the altitude of 2000 km (————). The ionosphere is assumed to be at the daytime sunspot minimum condition (Hughes and Southwood, 1976b). Only Alfven-mode wave is assumed to be present above the ionosphere. Azimuth is measured clockwise from the north.

is not significantly rotated through the ionosphere. Thus further refinement seems to be needed before the model can be compared with the observation.

At this point a brief comment is in order concerning the nature of preliminary and following impulses of SI discussed in Section I.5. The time scale of the rise, or fall, of SI is around 1 min and the screening theory developed above is applicable to (and in fact developed initially for) the SI phenomenon. According to Nishida (1964b), the preliminary impulse of a positive SI is explained as follows. In the early stage of the magnetospheric compression, the deformation occurs preferentially on the dayside of the magnetosphere and it produces a flow of magnetospheric plasma away from the noon meridian (c.f. Fig. 14). The associated magnetic perturbation is incident on the ionosphere as the Alfven-mode wave. The magnetic component of this incident wave is directed toward the noon meridian in the northern hemisphere and away from it in the southern hemisphere. This is so because the high-altitude portion of field lines is pushed away from the sun while their root is anchored to the ionosphere. The ionospheric screening rotates the magnetic perturbation vector through 90°. Since the sense of the rotation is counterclockwise (when viewed toward the ground) in the northern hemisphere and clockwise in the southern hemisphere, the resulting perturbation field on the ground is a decrease in the horizontal component in the afternoon sector. That is, the preliminary impulse of a positive SI is produced. The preliminary impulse of a negative SI can be explained in a similar way. The rotation of the perturbation field vector is caused, of course, by the ionospheric Hall current, which is the same current as the one incorporated in Tamao's (1964) theory of the preliminary impulse. Thus the SI* theories of these two authors are different expressions of the same idea.

Ionospheric Waveguide

In the estimate of the ionospheric-atmospheric modification that is summarized by Eq. (168) or Figure 114, the magnetosphere was supposed to be uniform. This, of course, is an oversimplification, because in low altitudes V_A varies significantly with altitude due to the sharp variation of the ion density with increasing height. In particular, there is a minimum of V_A at the peak of the ionospheric F_2 layer, and it is conceivable that the wave energy is trapped in this layer and ducted horizontally as it does in the waveguide. The situation is depicted schematically in Figure 116 (Manchester, 1966; Tepley and Landshoff, 1966). The wave to be ducted should be the fast mode that can transport energy horizontally across the ambient magnetic field. Due to nonuniformity of the medium, however, this mode is coupled with the Alfven mode, and the coupling

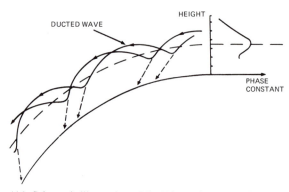

Fig. 116. Schematic illustration of the F-layer duct (Manchester, 1966)

acts as a sink of the trapped wave energy since the Alfven mode transports the energy parallel to the ambient magnetic field and hence away
from the trapping layer. The wave energy is lost also by the dissipation
in the ionosphere.

The horizontal phase velocity of the trapped wave, numerically derived, is given in Figure 117a (Greifinger and Greifinger, 1968). Roughly
speaking, the unit of the phase velocity in this Figure is the Alfven velocity
at its minimum, V_m, and that of the angular frequency is the ratio V_m/D
where D is the thickness of the low velocity layer. As the Figure shows,
the phase velocity, as well as the group velocity, is around V_m when ω
is sufficiently high. At lower ω, however, the phase velocity diverges and
there is a cutoff at $f_c \sim V_m/D$ for the lowest-order mode of the trapped
wave. Thus the ducting is limited to waves whose vertical wavelength in
the low V_A region is less than the thickness of that region. The cutoff
frequency is around 0.5 Hz. The group velocity tends toward zero at the
cutoff since

$$V_g = \frac{\omega/k}{1 - kd(\omega/k)/d\omega} \to 0 \quad \text{as} \quad \frac{d(\omega/k)}{d\omega} \to -\infty$$

The horizontal attenuation coefficient k_i is shown in Figure 117b in the
unit of $1/D$. The Figure shows the propagation in the geomagnetic meridian plane, and the decrease in the wave amplitude with horizontal distance becomes more pronounced for the off-meridional propagation due
to the strong coupling between the fast and the Alfven mode. Relatively
less attenuation is expected for the ducting in the nighttime ionosphere
than in the daytime one (Greifinger and Greifinger, 1968, 1973).

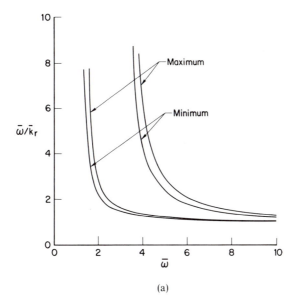

(a)

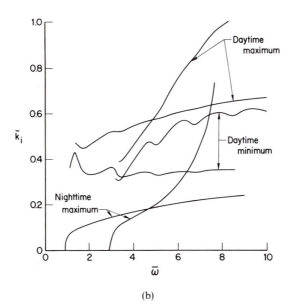

(b)

Fig. 117a,b. Phase velocity (a) and attenuation coefficient (b) of the ionospheric duct propagation for two lowest order modes of the trapped wave under different conditions of sunspot activity. For the unit of the variables, see text (Greifinger and Greifinger, 1968)

By virtue of the ducting the activity of the short-period pulsations can spread efficiently over the earth. The effect is demonstrated in Figure 118 where the sonagrams of pc1 waves at Kauai (22°) and Canton Island (−5°) are compared. The point of interest is that the repetition frequency of the fine structure at the near-equatorial station Canton is twice that at the northern low-latitude station Kauai. Since the fine structures of this type are known to appear alternately in northern and southern high latitudes, the 'structure doubling' at the near-equatorial station can be attributed to the superposition of the waves that have been ducted from higher northern and southern latitudes toward the equatorial region (Tepley, 1964). The structure doubling, however, is not a very common phenomenon, indicating that the wave energy tends to be lost before crossing the equator. In fact only 4% of the total hydromagnetic emissions at College (geomagnetic coordinates: 65°, 257°) are observed also at Great Whale River (67°, 348°) and Boulder (49°, 317°) (Fraser, 1975b). The attenuation in the duct seems to be less in the nighttime than in the daytime, since the occurrence frequency of pc1 waves in low latitudes is higher during the night than during the day, while in high latitudes pc1 waves are more frequent in the daytime (Manchester, 1966).

The horizontal group velocity of the pc1 waves has been derived from the comparison of the arrival times at a number of stations, and cases have

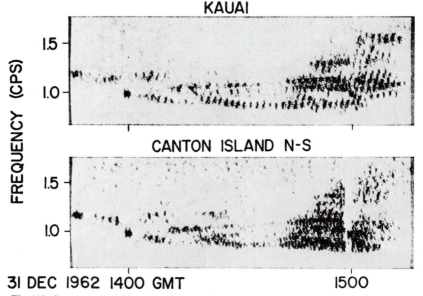

Fig. 118. Structure doubling of pc1 waves. The repetitive fine structure appears twice as often at Canton Island as at Kauai (Tepley, 1964)

been found where the velocity obtained agrees nicely with the value V_m expected for the ducted wave. In Figure 119 the observed group velocity is plotted against $(f_0F_2)^{-1}$, which is related to the electron density at the V_A minimum. The straight line gives the minimum Alfven speed V_m estimated on the assumption that the V_A minimum occurs at the altitude of 350 km, and good agreement is seen in many cases (Manchester, 1968; Fraser, 1975a). There are cases, however, where the observationally derived V_g is much greater than the estimated ducting speed, and cases have also been found where the dispersion of the fine structure elements of pc1 waves varies with horizontal distance (Campbell and Thornberry, 1972). The reason behind these observations remains to be found. As for the ellipticity of pc1 waves, the left-hand polarization is more frequent in high latitudes, but in low latitudes left-hand, linear, and right-hand polarizations are mixed (Kawamura, 1970). Since in the neutral atmosphere the $j_z = 0$ condition makes the perturbation magnetic field parallel, or antiparallel, to the wave vector in the horizontal plane, the major axis of the

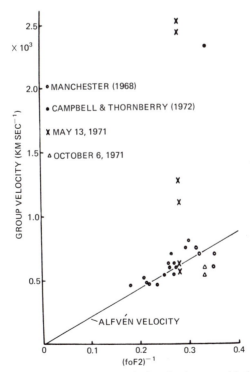

Fig. 119. Variation of the horizontal group velocity of pc1 waves with the electron density at the peak of the F_2 layer (Fraser, 1975a)

polarization ellipse gives a measure of the local direction of propagation of the wave (Summers, 1974). It is not yet clear, however, what is meant by elliptic polarizations that are observed in low latitudes with variable sense of polarization.

The ducting is not effective for waves having frequencies below the cut-off frequency of ~ 0.5 Hz. This means that among the family of magnetic pulsations only pc1 and part of pi1 can be guided through the waveguide located at the bottom of the magnetosphere. Thus for the majority of magnetic pulsations the energy constantly leaks across the ionosphere toward the magnetosphere at the same time as it propagates horizontally, and the wave energy is spread through the entire magnetosphere rather than being confined in the narrow ionospheric waveguide. Despite the resulting dilution of the wave energy, long-period pulsations of pc4 and pc5 types, whose resonant region is thought to be located at middle to high latitudes, have been observed even in the equatorial region, and the equatorial enhancement is noted at the day-side equator (e.g., Sarma et al., 1974; Jain and Srinivasacharya, 1975). The process of this propagation (particularly the process leading to the equatorial enhancement of the pc waves) is of much interest but is yet to be examined quantitatively. The major difficulty is that the ray theory cannot be applied to the horizontal propagation of these long-period waves.

References

Afonia, R. G., Feldstein, Y. I., Shabansky, V. P.: Aurora and Airglow **22**, 34 (1975)
Akasofu, S.-I.: Polar and Magnetospheric Substorms. Dordrecht: D. Reidel, 1968
Akasofu, S.-I.: Planet. Space Sci. **23**, 1349 (1975)
Akasofu, S.-I., Chapman, S.: Planet. Space Sci. **12**, 607 (1964)
Akasofu, S.-I., Snyder, A. L.: J. Geophys. Res. **77**, 6275 (1972)
Akasofu, S. I., Hones, E. W., Jr., Bame, S. J., Asbridge, J. R., Lui, A. T. Y.: J. Geophys. Res. 7257 (1973a)
Akasofu, S.-I., Perreault, P. D., Yasuhara, F., Meng, C.-I.: J. Geophys. Res. **78**, 7490 (1973b)
Akasofu, S.-I., Yasuhara, F., Kawasaki, K.: Planet. Space Sci. **21**, 2232 (1973c)
Alfven, H.: Kgl. Sv. Vet. Akad. Handl., Ser. 3, **18**, no. 3 (1939)
Alfven, H.: Tellus **7**, 50 (1955)
Angerami, J. J., Thomas, J. O.: J. Geophys. Res. **69**, 4537 (1964)
Araki, T.: Planet. Space Sci. **25**, 373 (1977)
Armstrong, J. C., Akasofu, S.-I., Rostoker, G.: J. Geophys. Res. **80**, 575 (1975)
Arnoldy, R. L.: J. Geophys. Res. **76**, 5189 (1971)
Arnoldy, R. L., Lewis, P. B., Isaacson, P. O.: J. Geophys. Res. **79**, 4208 (1974)
Atkinson, G.: J. Geophys. Res. **71**, 5157 (1966)
Atkinson, G.: J. Geophys. Res. **75**, 4746 (1970)
Aubry, M. P., Russell, C. T., Kivelson, M. G.: J. Geophys. Res. **75**, 7018 (1970)
Axford, W. I.: Rev. Geophys. **7**, 421 (1969)
Axford, W. I., Hines, C. O.: Can. J. Phys. **39**, 1433 (1961)
Baker, W. G., Martyn, D. F.: Phil. Trans. Roy. Soc. London, **A246**, 281 (1953)
Ballif, J. R., Jones, D. E., Coleman, P. J., Jr.: J. Geophys. Res. **74**, 2289 (1969)
Balsley, B. B.: J. Atmos. Terr. Phys. **35**, 1035 (1973)
Barfield, J. N., McPherron, R. L.: J. Geophys. Res. **77**, 4720 (1972)
Barfield, J. N., McPherron, R. L., Coleman, P. J., Jr., Southwood, D. J.: J. Geophys. Res. **77**, 143 (1972)
Bartels, J.: Ann. Int. Geophys. Year **4**, part 4, p. 127 (1957)
Behannon, K. W.: J. Geophys. Res. **75**, 743 (1970)
Belcher, J. W., Davis, L., Jr.: J. Geophys. Res. **76**, 3534 (1971)
Berko, F. W., Cahill, L. J., Jr., Fritz, T. A.: J. Geophys. Res. **80**, 3549 (1975)
Berthelier, A.: J. Geophys. Res. **81**, 4546 (1976)
Berthelier, A., Guerin, C.: Space Research XIII. Berlin: Akademie-Verlag, 1973, p. 661
Block, L. P.: J. Geophys. Res. **71**, 855 (1966)
Block, L. P., Fälthammar, C.-G.: Ann. Geophys. **32**, 161 (1976)
Boller, B. R., Stolov, H. L.: J. Geophys. Res. **75**, 6073 (1970)
Bolshakova, O. V., Troitskaya, V. A.: Dokl. Akad. Nauk USSR **180**, 343 (1968)
Boström, R.: J. Geophys. Res. **69**, 4983 (1964)
Brekke, A., Doupnik, J. R., Banks, P. M.: J. Geophys. Res. **79**, 3773 (1974)

Brown, R. R.: J. Geophys. Res. **78**, 5698 (1973)

Brown, W. L., Cahill, L. J., Davis, L. R., McIlwain, C. E., Roberts, C. S.: J. Geophys. Res. **73**, 153 (1968)

Burch, J. L.: J. Geophys. Res. **77**, 5629 (1972a)

Burch, J. L.: J. Geophys. Res. **77**, 6696 (1972b)

Burch, J. L.: Radio Sci. **8**, 955 (1973)

Burch, J. L., Lennartsson, W., Hanson, W. B., Heelis, R. A., Hoffman, J. H., Hoffman, R.A.: J. Geophys. Res. **81**, 3886 (1976)

Burke, W. J., Reasoner, D. L.: J. Geophys. Res. **78**, 6790 (1973)

Burlaga, L. F.: In: Solar-terrestrial Physics. Part II, Dyer, E. R. (ed.), Dordrecht; D. Reidel, 1970, p. 135

Burton, R. K., McPherron, R. L., Russell, C. T.: J. Geophys. Res. **80**, 4204 (1975)

Cahill, L. J., Jr.: J. Geophys. Res. **75**, 3778 (1970)

Cahill, L. J., Jr., Lee, Y. C.: Planet. Space Sci. **23**, 1279 (1975)

Camidge, F. P., Rostoker, G.: Can. J. Phys. **48**, 2002 (1970)

Campbell, W. H., Thornberry, T. C.: J. Geophys. Res. **77**, 1941 (1972)

Carpenter, D. L.: J. Geophys. Res. **71**, 693 (1966)

Carpenter, D. L., Akasofu, S.-I.: J. Geophys. Res. **77**, 6854 (1972)

Carpenter, D. L., Stone, K., Siren, J. C., Crystal, T. L.: J. Geophys. Res. **77**, 2819 (1972)

Chao, J. K., Olbert, S.: J. Geophys. Res. **75**, 6394 (1970)

Chappell, C. R.: In: Earth's Magnetospheric Processes. McCormac, B. M. (ed.), Dordrecht: D. Reidel, 1972, p. 280

Chappell, C. R., Harris, K. K., Sharp, G. W.: J. Geophys. Res. **75**, 50 (1970)

Chapman, S., Ferraro, V. C. A.: Terrest. Mag. Atmosph. Elec. **36**, 77 (1931)

Chapman, S., Bartels, J.: Geomagnetism. Oxford Univ. Press 1940

Chen, A. J., Wolf, R. A.: Planet. Space Sci. **20**, 483 (1972)

Choe, J. Y., Beard, D. B.: Planet. Space Sci. **22**, 595 (1974)

Choe, J. Y., Beard, D. B., Sullivan, E. C.: Planet. Space Sci. **21**, 485 (1973)

Cocke, W. J., Cornwall, J. M.: J. Geophys. Res. **72**, 2843 (1967)

Colburn, D. S., Sonett, C. P.: Space Sci. Rev. **5**, 439 (1966)

Cornwall, J. M.: J. Geophys. Res. **70**, 61 (1965)

Coroniti, F. V., Kennel, C. F.: J. Geophys. Res. **77**, 2835 (1972a)

Coroniti, F. V., Kennel, C. F.: J. Geophys. Res. **77**, 3361 (1972b)

Coroniti, F. V., Kennel, C. F.: J. Geophys. Res. **78**, 2837 (1973)

Cowley, S. W. H.: Cosm. Electrodyn. **3**, 448 (1973)

Cowley, S. W. H., Ashour-Abdalla, M.: Planet. Space Sci. **23**, 1527 (1975)

Crooker, N. U., McPherron, R. L.: J. Geophys. Res. **77**, 6886 (1972)

Davis, T. N., Sugiura, M.: J. Geophys. Res. **71**, 785 (1966)

Dessler, A. H., Francis, W. E., Parker, E. N.: J. Geophys. Res. **65**, 2715 (1960)

Dryer, M.: Space Sci. Rev. **17**, 277 (1975)

Dungey, J. W.: Penn. State Sci. Rep. No. 69 (1954)

Dungey, J. W.: In: Geophysics. DeWitt, C., et al. (ed.), New York: Gordon and Breach, 1962, p. 503

Ejiri, M., Hoffman, R. A., Smith, P. H.: J. Geophys. Res. **83**, in press (1978)

Fairfield, D. H.: J. Geophys. Res. **76**, 6700 (1971)

Fairfield, D. H.: Rev. Geophys. Space Phys. **14**, 117 (1976)

Fairfield, D. H., Cahill, L. J., Jr.: J. Geophys. Res. **71**, 155 (1966)

Fairfield, D. H., Ness, N. F.: J. Geophys. Res. **75**, 7032 (1970)

Fälthammar, C.-G.: Rev. Geophys. Space Phys.: **15**, 457 (1977)

Feldstein, Y. I.: Geomag. Aeron. **3**, 183 (1963)

Feldstein, Y. I.: J. Geophys. Res. **78**, 1210 (1973)
Feldstein, Y. I., Starkov, G. V.: Planet. Space Sci. **15**, 209 (1967)
Foster, J. C., Fairfield, D. H., Ogilvie, K. W., Rosenberg, T. J.: J. Geophys. Res. **76**, 6971 (1971)
Frank, L. A.: J. Geophys. Res. **76**, 2265 (1971)
Fraser, B. J.: J. Geophys. Res. **80**, 2790 (1975a)
Fraser, B. J.: J. Geophys. Res. **80**, 2797 (1975b)
Friis-Christensen, E., Wilhjelm, J.: J. Geophys. Res. **80**, 1248 (1975)
Friis-Christensen, E., Lassen, K., Wilhjelm, J., Wilcox, J. M., Gonzalez, W., Colburn, D. S.: J. Geophys. Res. **77**, 3371 (1972)
Fukunishi, H.: Rep. Ionos. Space Res. Jap. **23**, 21 (1969)
Fukunishi, H.: J. Geophys. Res. **80**, 98 (1975)
Fukunishi, H., Hirasawa, T.: Rep. Ionos. Space Res. Jap. **24**, 45 (1970)
Fukushima, N.: Rep. Ionos. Space Res. Jap. **23**, 219 (1969)
Fukushima, N.: preprint, GRL. Univ. Tokyo (1976)
Gendrin, R., Lacourly, S., Roux, A., Solomon, J., Feïgin, F. Z., Gokhberg, M. V., Troitskaya, V. A., Yakimenko, V. L.: Planet. Space Sci. **19**, 165 (1971)
Gonzalez, W. D., Mozer, F. S.: J. Geophys. Res. **79**, 4186 (1974)
Grebowsky, J. M., Chen, A. J.: Planet. Space Sci. **23**, 1045 (1975)
Greenstadt, E. W.: J. Geophys. Res. **77**, 1729 (1972)
Greifinger, C., Greifinger, P. S.: J. Geophys. Res. **73**, 7473 (1968)
Greifinger, C., Greifinger, P. S.: J. Geophys. Res., **78**, 4611 (1973)
Gul'elmi, A. V.: Space Sci. Rev. **16**, 331 (1974)
Gul'elmi, A. V., Bol'shakova, O. V.: Geomagn. Aeron. **13**, 459 (1973)
Gul'elmi, A. V., Plyasova-Bakunina, T. A., Shehepetnov, R. V.: Geomagn. Aeron. **13**, 331 (1973)
Gurnett, D. A.: Critical Problems of Magnetospheric Physics. Dryer, E. R. (ed.), Washington D.C.: NAS, 1972, p. 123
Gurnett, D. A., Frank, L. A.: J. Geophys. Res. **78**, 145 (1973)
Haerendel, G., Lüst, R.: Earth's Particles and Fields. McCormac, B. M. (ed.), New York-Amsterdam-London: Reinhold, 1968, p. 271
Hakura, Y.: Rep. Ionos. Space Res. Jap. **19**, 121 (1965)
Hasegawa, A.: Phys. Fluids **12**, 2642 (1969)
Hasegawa, A., Chen, L.: Space Sci. Rev. **16**, 347 (1974)
Hayashi, K., Kokubun, S., Oguti, T.: Rep. Ionos. Space Res. Jap. **22**, 149 (1968)
Heacock, R. R.: J. Geophys. Res. **76**, 4494 (1971)
Heacock, R. R., Kivinen, M.: J. Geophys. Res. **77**, 6764 (1972)
Heacock, R. R., Akasofu, S.-I.: J. Geophys. Res. **78**, 5524 (1973)
Heacock, R. R., Henderson, D. J., Reid, J. S., Kivinen, M. L: J. Geophys. Res. **81**, 273 (1976)
Heikkila, W. J., Winningham, J. D., Eather, R. H., Akasofu, S.-I.: J. Geophys. Res. **77**, 4100 (1972)
Heppner, J. P.: J. Geophys. Res. **77**, 4877 (1972)
Hines, C. O.: Space Sci. Rev. **3**, 342 (1964)
Hirasawa, T.: Rep. Ionos. Space Res. Jap. **23**, 281 (1969)
Hirasawa, T.: Rep. Ionos. Space Res. Jap. **24**, 66 (1970)
Hoffman, R. A., Bracken, P. A.: J. Geophys. Res. **72**, 6039 (1967)
Hoffman, R. A., Burch, J. L.: J. Geophys. Res. **78**, 2867 (1973)
Holzer, T. E., Reid, G. C.: J. Geophys. Res. **80**, 2041 (1975)
Hones, E. W., Jr.: Solar Phys. **47**, 101 (1976)

Hones, E. W., Jr.: J. Geophys. Res. **82**, 5633 (1977)

Hones, E. W., Jr., Akasofu, S.-I., Perreault, P., Bame, S. J., Singer, S.: J. Geophys. Res. **75**, 7060 (1970)

Hones, E. W., Jr., Singer, S., Lanzerotti, L. J., Pierson, J. D., Rosenberg, T. J.: J. Geophys. Res. **76**, 2977 (1971a)

Hones, E. W., Jr., Akasofu, S.-I., Bame, S. J., Singer, S.: J. Geophys. Res. **76**, 8241 (1971b)

Hones, E. W., Jr., Asbridge, J. R., Bame, S. J., Singer, S.: J. Geophys. Res. **78**, 109 (1973)

Hones, E. W., Jr., Lui, A. T. Y., Bame, S. J., Singer, S.: J. Geophys. Res. **79**, 1385 (1974)

Horning, B. L., McPherron, R. L., Jackson, D. D.: J. Geophys. Res. **79**, 5202 (1974)

Hughes, W. J., Southwood, D. J.: J. Geophys. Res. **81**, 3234 (1976a)

Hughes, W. J., Southwood, D. J.: J. Geophys. Res. **81**, 3241 (1976b)

Hundhausen, A. J.: Solar Wind and Coronal Expansion. Berlin-Heidelberg-New York: Springer Verlag 1972

Iijima, T., Nagata, T.: Rep. Ionos. Space Res. Jap. **22**, 1 (1968)

Iijima, T., Nagata, T.: Planet. Space Sci. **20**, 1095 (1972)

Iijima, T., Potemra, T. A.: J. Geophys. Res. **81**, 2165 (1976a)

Iijima, T., Potemra, T. A.: J. Geophys. Res. **81**, 5971 (1976b)

Iwasaki, N.: Rep. Ionos. Space Res. Jap. **25**, 163 (1971)

Jacobs, J. A.: Geomagnetic Micropulsations. Berlin-Heidelberg-New York: Springer-Verlag 1970

Jacobs, J. A., Watanabe, T.: J. Atmosph. Terrest. Phys. **26**, 825 (1964)

Jaggi, R. K., Wolf, R. A.: J. Geophys. Res. **78**, 2852 (1973)

Jain, A. R., Srinivasacharya, K. G.: J. Atmos. Terr. Phys. **37**, 1477 (1975)

Jensen, D. C., Cain, J. C.: J. Geophys. Res. **67**, 3568 (1962)

Kamide, Y., Fukushima, N.: Rep. Ionos. Space Res. Jap. **25**, 125 (1971)

Kamide, Y., Fukushima, N.: Rep. Ionos. Space Res. Jap. **26**, 79 (1972)

Kamide, Y., McIlwain, C. E.: J. Geophys. Res. **79**, 4787 (1974)

Kamide, Y., Akasofu, S.-I.: J. Geophys. Res. **79**, 3755 (1974)

Kamide, Y., Brekke, A.: J. Geophys. Res. **80**, 587 (1975)

Kamide, Y., Akasofu, S.-I., Deforest, S. E., Kisabeth, J. L.: Planet. Space Sci. **23**, 579 (1975)

Kamide, Y., Yasuhara, F., Akasofu, S.-I.: Planet. Space Sci. **24**, 215 (1976)

Kamide, Y., Perreault, P. D., Akasofu, S.-I., Winningham, J. D.: J. Geophys. Res. **82**, 5521 (1977)

Kavanagh, L. D., Jr., Freeman, J. W., Jr. Chen, A. J.: J. Geophys. Res. **73**, 5511 (1968)

Kawamura, M.: Memoirs of the Kakioka Magnetic Observatory. Suppl. **3**, 1970

Kawasaki, K., Akasofu, S.-I.: J. Geophys. Res. **72**, 5363 (1967)

Kawasaki, K., Akasofu, S.-I.: Planet. Space Sci. **19**, 1339 (1971a)

Kawasaki, K., Akasofu, S.-I.: J. Geophys. Res. **76**, 2396 (1971b)

Kawasaki, K., Akasofu, S.-I.: Planet. Space Sci. **20**, 1163 (1972)

Kawasaki, K., Akasofu, S.-I.: Planet. Space Sci. **21**, 329 (1973)

Kawasaki, K., Akasofu, S.-I., Yasuhara, F., Meng, C.-I.: J. Geophys. Res. **76**, 6781 (1971)

Kawasaki, K., Yasuhara, F., Akasofu, S.-I.: Planet. Space Sci. **21**, 1743 (1973)

Keath, E. P., Roelof, E. C., Bostrom, C. O., Williams, D. J.: J. Geophys. Res. **81**, 2315 (1976)

Kennel, C. F., Petschek, H. E.: J. Geophys. Res. **71**, 1 (1966)

Kenney, J. F., Knaflich, H. B., Liemohn, H. B.: J. Geophys. Res. **73**, 6737 (1968)

Kimura, I., Matsumoto, H.: Radio Sci. **3**, 333 (1968)

Kisabeth, J. L., Rostoker, G.: J. Geophys. Res. **79**, 972 (1974)

Kokubun, S.: Rep. Ionos. Space Res. Jap. **24**, 24 (1970)

Kokubun, S.: Planet. Space Sci. **19**, 697 (1971)

Kokubun, S., Oguti, T.: Rep. Ionos. Space Res. Jap. **22**, 45 (1968)
Kokubun, S., Iijima, T.: Planet. Space Sci. **23**, 1483 (1975)
Kokubun, S., McPherron, R. L., Russell, C. T.: J. Geophys. Res. **81**, 5141 (1976)
Kokubun, S., Kivelson, M. G., McPherron, R. L., Russell, C. T., West, H. I., Jr.: J. Geophys. Res. **81**, 2774 (1977)
Kokubun, S., McPherron, R. L., Russell, C. T.: J. Geophys. Res. **82**, 74 (1977)
Kosik, J. C.: Ann. Geophys. **27**, 11 (1971)
Langel, R. A.: Planet. Space Sci. **22**, 1413 (1974)
Langel, R. A., Sweeney, R. E.: J. Geophys. Res. **76**, 4420 (1971)
Lanzerotti, L. J., Fukunishi, H.: J. Geophys. Res. **80**, 4627 (1975)
Lanzerotti, L. J., Hasegawa, A., Maclennan, C. G.: J. Geophys. Res. **74**, 5565 (1969)
Lanzerotti, L. J., Fukunishi, H., Chen, L.: J. Geophys. Res. **79**, 4648 (1974)
Lanzerotti, L. J., Maclennan, C. G., Fukunishi, H.: J. Geophys. Res. **81**, 3299 (1976)
Lezniak, T. W., Winckler, J. R.: J. Geophys. Res. **75**, 7075 (1970)
Lui, A. T. Y., Anger, C. D., Venkatesan, D., Sawchuk, W., Akasofu, S.-I.: J. Geophys. Res. **80**, 1795 (1975a)
Lui, A. T. Y., Anger, C. D., Akasofu, S.-I.: J. Geophys. Res. **80**, 3603 (1975b)
Lui, A. T. Y., Hones, E. W., Jr., Venkatesan, D., Akasofu, S.-I., Bame, S. J.: J. Geophys. Res. **80**, 4649 (1975c)
Lui, A. T. Y., Meng, C.-I., Akasofu, S.-I.: J. Geophys. Res. **82**, 1547 (1977)
Lyatsky, W. B., Maltsev, Yu. P., Leontyev, S. V.: Planet. Space Sci. **22**, 1231 (1974)
Maeda, H.: J. Geomag. Geoelect. **7**, 121 (1955)
Maezawa, K.: Planet. Space Sci. **22**, 1443 (1974)
Maezawa, K.: J. Geophys. Res. **80**, 3543 (1975)
Maezawa, K.: J. Geophys. Res. **81**, 2289 (1976)
Maezawa, K., Nishida, A.: J. Geomag. Geoelect. **29**, in press (1978)
Mansurov, S. M.: Geomag. Aeron. **4**, 622 (1969)
Manchester, R. N.: J. Geophys. Res. **71**, 3749 (1966)
Manchester, R. N.: J. Geophys. Res. **73**, 3549 (1968)
Matsushita, S.: J. Geophys. Res. **67**, 3753 (1962)
Matsushita, S.: J. Geophys. Res. **70**, 4395 (1965)
Matsushita, S., Maeda, H.: J. Geophys. Res. **70**, 2559 (1965)
Matsushita, S., Balsley, B. B.: Planet. Space Sci. **20**, 1259 (1972)
Mayaud, P. N.: Ann. Geophys. **23**, 585 (1967)
Mayaud, P. N.: J. Geophys. Res. **80**, 111 (1975)
Maynard, N. C.: J. Geophys. Res. **79**, 4620 (1974)
Maynard, N. C., Chen, A. J.: J. Geophys. Res. **80**, 1009 (1975)
McClay, J. F.: Planet. Space Sci. **21**, 2193 (1973)
McDiarmid, I. B., Burrows, J. R., Budzinski, E. E.: J. Geophys. Res. **81**, 221 (1976)
McIlwain, C. E.: J. Geophys. Res. **66**, 3681 (1961)
McKenzie, J. F.: Planet. Space Sci. **18**, 1 (1970)
McPherron, R. L.: J. Geophys. Res. **75**, 5592 (1970)
McPherron, R. L.: Planet. Space Sci. **20**, 1521 (1972)
McPherron, R. L., Aubry, M. P., Russell, C. T., Coleman, P. J., Jr.: J. Geophys. Res. **78**, 3068 (1973a)
McPherron, R. L., Russell, C. T., Aubry, M. P.: J. Geophys. Res. **78**, 3131 (1973b)
Mead, G. D., Beard, D. B.: J. Geophys. Res. **69**, 1169 (1964)
Mende, S. B., Sharp, R. D., Shelley, E. G., Haerendel, G., Hones, E. W., Jr.: J. Geophys. Res. **77**, 4682 (1972)
Meng, C.-I., Akasofu, S.-I.: J. Geophys. Res. **72**, 4905 (1967)

Mihalov, J. D., Colburn, D. S., Sonett, C. P.: Planet. Space Sci. **18**, 239 (1970)
Mozer, F. S.: J. Geophys. Res. **76**, 7595 (1971)
Mozer, F. S., Gonzalez, W. D., Bogott, F., Kelley, M. C., Schutz, S.: J. Geophys. Res. **79**, 56 (1974)
Murayama, T.: J. Geophys. Res. **71**, 5547 (1966)
Murayama, T.: J. Geophys. Res. **79**, 297 (1974)
Murayama, T., Hakamada, K.: Planet. Space Sci. **23**, 75 (1975)
Nagata, T., Kokubun, S.: Rep. Ionos. Space Res. Jap. **16**, 256 (1962)
Nishida, A.: J. Geophys. Res. **69**, 947 (1964a)
Nishida, A.: J. Geophys. Res. **69**, 1861 (1964b)
Nishida, A.: Rep. Ionos. Space Res. Jap. **20**, 36 (1966a)
Nishida, A.: Rep. Ionos. Space Res. Jap. **20**, 42 (1966b)
Nishida, A.: J. Geophys. Res. **71**, 5669 (1966c)
Nishida, A.: J. Geophys. Res. **73**, 1795 (1968)
Nishida, A.: Planet. Space Sci. **19**, 205 (1971a)
Nishida, A.: Cosm. Electrodyn. **2**, 350 (1971b)
Nishida, A.: Planet. Space Sci. **21**, 1255 (1973)
Nishida, A., Jacobs, J. A.: J. Geophys. Res. **67**, 525 (1962a)
Nishida, A., Jacobs, J. A.: J. Geophys. Res. **67**, 4937 (1962b)
Nishida, A., Kokubun, S.: Rev. Geophys. Space Phys. **9**, 417 (1971)
Nishida, A., Maezawa, K.: J. Geophys. Res. **76**, 2254 (1971)
Nishida, A., Nagayama, N.: J. Geophys. Res. **78**, 3782 (1973)
Nishida, A., Hones, E. W., Jr.: J. Geophys. Res. **79**, 535 (1974)
Nishida, A., Nagayama, N.: Planet. Space Sci. **23**, 1119 (1975)
Northrop, T. G.: The Adiabatic Motion of Charged Particles. New York-London-Sydney: Interscience Publishers 1963
Nourrey, G. R.: Interplanetary magnetic field, solar wind and geomagnetic micropulsation, PhD thesis, Univ. British Columbia 1976
Obayashi, T.: J. Geophys. Res. **70**, 1069 (1965)
Obayashi, T., Jacobs, J. A.: Geophys. J. **1**, 53 (1958)
Ogilvie, K. W., Burlaga, L. F.: J. Geophys. Res. **79**, 2324 (1974)
Ogilvie, K. W., Burlaga, L. F., Wilkerson, T. D.: J. Geophys. Res. **73**, 6809 (1968)
Oguti, T., Watanabe, T.: J. Atmosph. Terrest. Phys. **38**, 543 (1976)
Olson, W. P.: J. Geophys. Res. **74**, 5642 (1969)
Olson, W. P.: J. Geophys. Res. **75**, 7244 (1970)
Olson, J. V., Rostoker, G.: Planet. Space Sci. **23**, 1129 (1975)
Orr, D.: J. Atmosph. Terrest. Phys. **35**, 1 (1973)
Orr, D., Webb, D. C.: Planet. Space Sci. **23**, 1169 (1975)
Park, C. G.: J. Geophys. Res. **75**, 4249 (1970)
Paschmann, G., Haerendel, G., Sckopke, N., Rosenbauer, H.: J. Geophys. Res. **81**, 2883 (1976)
Patel, V. L., Coleman, P. J., Jr.: J. Geophys. Res. **75**, 7255 (1970)
Patel, V. L., Cahill, L. J., Jr.: Planet. Space Sci. **22**, 1117 (1974)
Petschek, H. E.: NASA Spec. Publ. SP-50 (1964)
Petschek, H. E.: The Solar Wind. Mackin, R. J., Jr., Neugebauer, M. (eds.), Oxford: Pergamon Press, 1966, p. 257
Pike, C. P., Meng, C.-I., Akasofu, S.-I., Whalen, J. A.: J. Geophys. Res. **79**, 5129 (1974)
Pudovkin, M. I., Shumilov, O. I., Zaitzeva, S. A.: Planet. Space Sci. **16**, 881 (1968)
Rastogi, R. G.: J. Atmosph. Terrest. Phys. **24**, 1031 (1962)
Rastogi, R. G., Sastri, N. S.: J. Geomag. Geoelect. **26**, 529 (1974)
Rees, M. H., Roble, R. G.: Rev. Geophys. Space Phys. **13**, 201 (1975)

Reiff, P. H., Hill, T. W., Burch, J. L.: J. Geophys. Res. **82**, 479 (1977)

Rich, F. J., Reasoner, D. L., Burke, W. J.: J. Geophys. Res. **78**, 8097 (1973)

Richmond, A. D.: J. Geophys. Res. **81**, 1447 (1976)

Richmond, A. D., Matsushita, S., Tarpley, J. D.: J. Geophys. Res. **81**, 547 (1976)

Roederer, J. G.: Dynamics of Geomagnetically Trapped Radiation. Berlin–Heidelberg–New York: Springer Verlag 1970

Roederer, J. G., Hones, E. W., Jr.: J. Geophys. Res. **79**, 1432 (1974)

Rosenbauer, H., Grunwaldt, H., Montgomery, M. D., Paschmann, G., Sckopke, N.: J. Geophys. Res. **80**, 2723 (1975)

Rostoker, G.: Rev. Geophys. Space Phys. **10**, 935 (1972)

Rostoker, G., Fälthammar, C.-G.: J. Geophys. Res. **72**, 5853 (1967)

Rostoker, G., Kisabeth, J. L.: J. Geophys. Res. **78**, 5559 (1973)

Rostoker, G., Chen, A. J., Yasuhara, F., Akasofu, S.-I., Kawasaki, K.: Planet. Space Sci. **22**, 427 (1974)

Rostoker, G., Armstrong, J. C., Zmuda, A. J.: J. Geophys. Res. **80**, 3571 (1975)

Roth, B., Orr, D.: Planet. Space Sci. **23**, 993 (1975)

Russell, C. T.: In: Critical Problems of Magnetospheric Physics. Dryer, E. R. (ed.), Washington D.C.: NAS, 1972, p. 1

Russell, C. T.: In: Correlated Interplanetary and Magnetospheric Observations. Page, D. E. (ed.), Dordrecht-Boston: D. Reidel, 1974, p. 3

Russell, C. T., Brody, K. I.: J. Geophys. Res. **72**, 6104 (1967)

Russell, C. T., McPherron, R. L.: J. Geophys. Res. **78**, 92 (1973a)

Russell, C. T., McPherron, R. L.: Space Sci. Rev. **15**, 205 (1973b)

Saito, T.: Space Sci. Rev. **10**, 319 (1969)

Saito, T.: Proc. Magnetosph. Symp., ISAS, Univ. Tokyo, 1976, p. 70

Saito, T., Matsushita, S.: Planet. Space Sci. **15**, 573 (1967)

Saito, T., Takahashi, F., Morioka, A., Kuwashima, M.: Planet. Space Sci. **22**, 939 (1974)

Saito, T., Sakurai, T., Koyama, Y.: J. Atmosph. Terrest. Phys. **38**, 1265 (1976)

Sakurai, T., Saito, T.: Planet. Space Sci. **24**, 573 (1976)

Samson, J. C.: J. Geophys. Res. **77**, 6145 (1972)

Samson, J. C., Rostoker, G.: J. Geophys. Res. **77**, 6133 (1972)

Sarma, S. V. S., Ramanujachang, K. R., Narayan, P. V. S.: Indian J. Radio Space Phys. **3**, 221 (1974)

Sarris, E. T., Krimigis, S. M., Armstrong, T. P.: J. Geophys. Res. **81**, 2341 (1976)

Sato, T.: preprint, GRL, Univ. Tokyo (1976)

Schindler, K.: J. Geophys. Res. **79**, 2803 (1974)

Schulz, M., Lanzerotti, L. J.: Particle Diffusion in the Radiation Belts. Berlin-Heidelberg-New York: Springer Verlag 1974

Sckopke, N.: J. Geophys. Res. **71**, 3125 (1966)

Sckopke, N., Paschmann, G., Rosenbauer, H., Fairfield, D. H.: J. Geophys. Res. **81**, 2687 (1976)

Shelley, E. G., Sharp, R. D., Johnson, R. G.: J. Geophys. Res. **81**, 2363 (1976)

Singer, S.: Trans. Am. Geophys. Union **38**, 175 (1957)

Siscoe, G. L.: J. Geophys. Res. **75**, 5340 (1970)

Siscoe, G. L., Crooker, N. U.: J. Geophys. Res. **79**, 1110 (1974a)

Siscoe, G. L., Crooker, N. U.: Geophys. Res. Lett. **1**, 17 (1974b)

Siscoe, G. L., Formisano, V., Lazarus, A. J.: J. Geophys. Res. **73**, 4869 (1968)

Smith, P. H., Hoffman, R. A.: J. Geophys. Res. **79**, 966 (1974)

Smith, P. H., Hoffman, R. A., Fritz, T. A.: J. Geophys. Res. **81**, 2701 (1976)

Snyder, A. L., Akasofu, S.-I.: J. Geophys. Res. **77**, 3419 (1972)

Sonnerup, B. U. O.: J. Geophys. Res. **79**, 1546 (1974)

Sonnerup, B. U. O., Ledley, B. G.: J. Geophys. Res. **79**, 4309 (1974)

Southwood, D. J.: Planet. Space Sci. **16**, 587 (1968)

Southwood, D. J.: Planet. Space Sci. **22**, 483 (1974)

Spreiter, J. R., Summers, A. L., Alksne, A. Y.: Planet. Space Sci. **14**, 223 (1966)

Stix, T. H.: The Theory of Plasma Waves. New York: McGraw-Hill 1962

Sugiura, M.: Ann. Int. Geophys. Year **35**, 9 (1964)

Sugiura, M.: J. Geophys. Res. **80**, 2057 (1975)

Sugiura, M., Heppner, J. P.: In: Introduction to Space Science, Hess, W. N. (ed.), New York-London-Paris: Gordon and Breach Science, 1965, p. 5

Sugiura, M., Skillman, T. L., Ledley, B. G., Heppner, J. P.: J. Geophys. Res. **73**, 6699 (1968)

Summers, W. R.: Planet. Space Sci. **22**, 801 (1974)

Svalgaard, L.: Dan. Meteorol. Inst. Geophys. Pap. R-6,1 (1968)

Svalgaard, L.: J. Geophys. Res. **78**, 2064 (1973)

Svalgaard, L.: J. Geophys. Res. **80**, 2717 (1975)

Tamao, T.: Rep. Ionos. Space Res. Jap. **18**, 16 (1964)

Tamao, T.: Phys. Fluids **12**, 1458 (1969)

Tepley, L.: J. Geophys. Res. **69**, 2273 (1964)

Tepley, L., Landshoff, R. K.: J. Geophys. Res. **71**, 1499 (1966)

Terasawa, T., Nishida, A.: Planet. Space Sci. **24**, 855 (1976)

Troitskaya, V. A., Gul'elmi, A. V.: Space Sci. Rev. **7**, 689 (1967)

Troitskaya, V. A., Shchepetnov, R. V., Gul'elmi, A. V.: Geomag. Aeron. **8**, 634 (1968)

Troshichev. O. A., Kuznetsov, B. M., Pudovkin, M. I.: Planet. Space Sci. **22**, 1403 (1974)

Van Sabben, D.: J. Atmosph. Terrest. Phys. **30**, 1641 (1968)

Vasyliunas, V. M.: J. Geophys. Res. **73**, 2839 (1968)

Wagner, C.-U.: J. Atmosph. Terrest. Phys. **33**, 751 (1971)

Walker, R. C., Villante, U., Lazarus, A. J.: J. Geophys. Res. **80**, 1238 (1975)

Wescott, E. M., Stolarik, J. D., Heppner, J. P.: J. Geophys. Res. **74**, 3469 (1969)

Wiens, R. G., Rostoker, G.: J. Geophys. Res. **80**, 2109 (1975)

Wilcox, J. M., Ness, N. F.: J. Geophys. Res. **70**, 5793 (1965)

Wilcox, J. M., Svalgaard, L., Hedgecock, P. C.: J. Geophys. Res. **80**, 3685 (1975)

Willis, D. M.: J. Atmosph. Terrest. Phys. **26**, 581 (1964)

Wilson, C. R., Sugiura, M.: J. Geophys. Res. **66**, 4097 (1961)

Winningham, J. D., Yasuhara, F., Akasofu, S.-I., Heikkila, W. J.: J. Geophys. Res. **80**, 3148 (1975)

Wolf, R. A.: In: Magnetospheric Physics, McCormac, B. M. (ed.). Dordrecht-Boston: D. Reidel, 1974, p. 167

Wolfe, A., Kaufmann, R. L.: J. Geophys. Res. **80**, 1764 (1975)

Yabuzaki, T., Ogawa, T.: J. Geophys. Res. **79**, 1999 (1974)

Yasuhara, F., Kamide, Y., Akasofu, S.-I.: Planet. Space Sci. **23**, 1355 (1975)

Subject Index

VANIER LIBRARY
UNIVERSITY OF